That's Not in My Geography Book

That's Not in My Geography Book

A Compilation of Little-Known Facts

Kate Kelly

Taylor Trade Publishing

Lanham • New York • Boulder • Toronto • Plymouth, UK

Published by Taylor Trade Publishing
An imprint of The Rowman & Littlefield Publishing Group, Inc.
4501 Forbes Boulevard, Suite 200, Lanham, Maryland 20706
www.rlpgtrade.com

Estover Road, Plymouth PL6 7PY, United Kingdom

Distributed by NATIONAL BOOK NETWORK

Library of Congress Cataloging-in-Publication Data
Kelly, Kate, 1950–
That's not in my geography book : a compilation of little-known facts / Kate Kelly.
p. cm.
Includes index.
ISBN-13: 978-1-58979-340-8 (pbk. : alk. paper)
ISBN-10: 1-58979-340-4 (pbk. : alk. paper)
ISBN-13: 978-1-58979-434-4 (electronic)
ISBN-10: 1-58979-434-6 (electronic)
1. Geography—Miscellanea. I. Title.
G131.K45 2009
910—dc22

2008047527

∞™ The paper used in this publication meets the minimum requirements of American National Standard for Information Sciences—Permanence of Paper for Printed Library Materials, ANSI/NISO Z39.48-1992.
Manufactured in the United States of America.

Contents

Contents

Acknowledgments

This book was made possible because of many hardworking individuals who have committed to making geography interesting and accessible to more people. I would like to express appreciation to Rick Rinehart, whose faith in me was unflagging; I thank him for his patience and support. Thank you, too, to Dulcie Wilcox and Janice Braunstein at Rowman & Littlefield, and to Catherine Bielitz. All of them contributed to making this book better. Kalen Landow has enthusiastically pitched in to help spread the word about the book, and I am grateful for the help.

Before the manuscript was delivered I benefited from the guidance and encouragement of agent Bob Diforio and researcher Bernadette Sukley, who was a gift in helping pull so much material together and make sense of it.

I hope book buyers will settle in to read the book with a map nearby. Our world is amazing, and you'll soon see why.

Introduction

> A knowledge of the Globe and its various inhabitants . . . never fails, in uncorrupted minds, to weaken local prejudices, and enlarge the sphere of benevolent feelings.
>
> —James Madison

Generally speaking, "geography" is the study of the earth and its features, its inhabitants, and the effects of human activity. As I struggled to define what should be in a book called *That's Not in My Geography Book*, I fought against the boring class about topography and population density that I remembered from junior high. One of my daughters loved the "where in the world" aspect of preparing for the school geography bee, and we spent hours of quality time running practice questions about capital cities and country locations as well as cultural geography questions about languages spoken and types of money exchanged. But that sort of information, too, did not feel like the right material for this book. Other than Amanda (the aforementioned daughter), who would think it was fun to learn names and dates and places?

What most people want when they pick up a book like this is a quick way to relate to the subject under investigation. And, of course, the best way to convey information is to find interesting facts that tell a story—in this case, a very important story about the geography of the planet on which we live. As we conclude the first decade of the twenty-first century, we face a world that offers both limits and opportunities. Marco Polo, Christopher Columbus, and Lewis and Clark had huge amounts of land to explore. We now have just a few remaining areas that are only somewhat unexplored, and yet our literacy

rate when it comes to geography is quite low. A poll taken in 2006 showed that after three years of battle in Iraq—and at that time, nearly 2,400 U.S. military deaths—almost two-thirds of Americans aged 18–24 still could not find Iraq on a map. And according to this study, the National Geographic–Roper Public Affairs 2006 Geographic Literacy Study, less than six months after Hurricane Katrina devastated New Orleans and the Gulf Coast, 33 percent of those polled could not point out Louisiana on a U.S. map. The study reported that "half or fewer of young men and women 18–24 can identify the states of New York or Ohio on a map [50 percent and 43 percent, respectively]" and concluded that the young people in the United States ". . . are unprepared for an increasingly global future." This point was brought home again during the 2008 Summer Olympics when we heard the news that Russia had invaded Georgia. Of course, the invasion was of Russia's neighboring country of Georgia, not the southern U.S. state of Georgia, though the news left a good number of people scratching their heads, unaware that there was a country named "Georgia." People need to know—and be interested in—where things are.

And that's how I arrived at what to put in this particular geography book. I found that the stories that kept leading me back to maps were some of the same stories of explorers I had read about fleetingly in school many years ago. However, today new research has provided scholars with a more accurate picture than the "myths" we were often told about men such as Columbus (many other people besides Columbus also knew the world was not flat) and Magellan (he personally never circumnavigated the globe), so I went back and began reading the stories of the men—and a few women—who really began to explore the geography of our globe. In many cases, I found stories about the people that are not commonly told, and I was always in awe of their bravery and nerve in venturing out on these trips to "who knows where," often guided by little more than the stars. As I read, I found the world "came alive" for me, and I remembered where things were so much more vividly because I understood how hard it was for the original explorers to get there.

My intention is to accomplish this same thing for readers—to send you off to check maps to find a location and then to be able to remember it because the story is so much clearer now. There are also some new details about recent archaeological finds that are quite fascinating and add factual details to the information scholars now have about these early explorations.

Of course, a book on geography would not be complete without some information on plate tectonics and how this works, as well as information about navigational methods and maps. Whether you opt for checking out maps on the computer, flipping through an atlas, or standing in front of a globe, there are few things more fascinating than a map because each one tells a different story, depending on the subject and the perspective of the map creator. From children who love to study Pooh's "100 Akre Wood" to locate Eeyore's Gloomy Place or Kanga's House to people who are fascinated by aeronautical charts for flight planning, maps tell many stories and will always be a vital part of our understanding of the world.

The book concludes with some thoughts on preserving our world. After reading Jared Diamond's 2005 book, *Collapse*, about how civilizations sometimes "do themselves in" through poor environmental management, I thought it was important to point readers toward this important work. The final chapter also focuses on water conservation and planning. I grew up in the arid West, where "liquor is for drinking and water is for fighting over." Though the surface of the earth contains much more water than land, less than 1 percent of the water supply can be used as drinking water, so I thought it important to remind people that no matter how hard it rains during any given year, we still have only a limited amount of this vital resource.

PART ONE

Discovering the World

1

The Very Early Adventurers

Including Forgotten Leif and Marco the Mythmaker

What's down this trail, around the next bend, or across the sea? Explorers from the early Phoenicians to the adventurers of today have been driven in pursuit of food, water, trade, the desire to conquer, or simply "the need to know." Long ago, maps were rudimentary or nonexistent, and these early travelers often packed for journeys that were expected to last for several years. They sailed uncharted waters to unknown lands with no particular idea of how or when they would come back. Bravery and a sense of adventure had to be their constant companions.

There are many misunderstandings about the early explorers who have shaped our view of the world. Either information about them was not written down during their time (as was the case with Leif Eriksson), or multiple hand-transcribed copies of a document were made or translated in the days before the printing press, leaving scholars today wondering what was accurate and what was added by the scribe or translator. As time has passed, more information has been uncovered—both in the form of additional printed versions of a story, but also in terms of archaeological proof of the various explorers' travels. In addition, technology has created enhanced capabilities to date materials, and as a result, scientists and historians have been better prepared to provide more complete information on the brave people who opened up the geography of our world.

These first chapters are to set right some of the information about who explored what, when, and how. Their stories are the ones that bring geography to life. It is hard to focus on a landmass without knowing about the personalities of the people who "discovered" it.

Who Were Really among the First?

Phoenicia was an ancient maritime country of southwest Asia consisting of city-states along the eastern Mediterranean Sea in present-day Syria and Lebanon. Its culture expanded across the Mediterranean between 1550 B.C.E. and 300 B.C.E. Its people became the foremost navigators and traders of the Mediterranean, establishing numerous colonies, including Carthage in northern Africa. The Phoenicians traveled to the edges of the known world at the time and were among the first to venture to the far western part of the Mediterranean and through the Strait of Gibraltar (then known as the "Pillars of Hercules"). The Phoenicians were in search of new sources of raw material and new markets for their products.

They traveled via a man-powered sailing vessel known as a galley, and they also invented a vessel known as the bireme, a more advanced form of galley with two tiers of oars, which would have increased speed. The people of this time had few navigational aids, so archaeologists think that the sailors traveled from village to village during the day, hugging the coastline as they went. Those who were traveling farther needed to be more adventurous and navigate in deeper waters, but they still kept within sight of land. Sailing at night was only done if they could keep tabs on the North Star, known then as the "Phoenician Star."

By the fifth century B.C.E., Himlico, an explorer from Carthage (a city settled by the Phoenicians), began to sail northward through the Strait of Gibraltar and around the western coasts of Spain and France. He traveled to the area the Phoenicians called "Tin-Land." At the time, the area (now France) was inhabited by people who controlled the tin trade. This greatly expanded the trade routes for these ancient people; they are also thought to have established settlements along the African coast. One of the other notable explorers of this time was Hanno, a Carthaginian who explored and colonized the west coast of Africa some time during the fifth century B.C.E. He established settlements in the area that is now Morocco and is thought to have made it as far as present-day Sierra Leone.

Leif Eriksson (fl. 1000 C.E.): An Undiscovered Explorer

While history and geography books now place Leif Eriksson in North America five hundred years before Columbus, this "update" only occurred in the 1960s. For almost a thousand years, Leif Eriksson's story was known only as a myth. The narrative was regularly shared with others as part of two Icelandic sagas that were told to pass the time during the long Icelandic nights. The sagas—*Eriks saga* and *Groenlendinga saga*—were viewed as marvelous adventures with no basis in truth. They were finally written down about 250 years after Leif lived. This delay heightened historians' skepticism about the information. Most maintained that the story was just a myth and that, of course, kept Leif out of the history and geography books. But Leif Eriksson's story eventually became known as an undiscovered part of our history when an archaeologist began poking around to see if he could find any truth in the tale. Here is what was learned.

During the height of the Middle Ages in Europe (ca. 980), Leif's father, Erik "the Red" Thorvaldson, a Norwegian, was exiled from Norway to Iceland because he was too violent for the Norse community. Iceland did not bring out any better behavior in their new resident; Erik killed a man there during a dispute. He was banished from Iceland for three years, but Erik had a plan. From other travelers passing through, Erik had heard about land west of Iceland, so he packed his boat and sailed west with his family. Though much of the new land they came upon was surrounded by glaciers and icebergs, there were also patches of green, so Erik called the land "Greenland," hoping to attract other people. Three years later, Erik returned to Iceland to tell others about the land he had found, and he organized twenty-five ships to bring settlers to his "Green Land." Bad weather interrupted the trip, and only fourteen ships with about 350 individuals arrived. Despite this difficult start, the community flourished for four hundred years, unusually long for a new settlement in a harsh environment.

Leif was the second of Erik's three sons. Though the exact year of his birth is uncertain, he was born while the family

A painting by Norwegian artist Christian Krohg shows Leif Eriksson steering his ship through dangerous waters. Source: Christian Krohg, Prints & Photographs Division, Library of Congress, LC-USZ62-3028

lived in Iceland, some time between 960 and 980 C.E. At about the age of twenty, he is thought to have returned to Norway. There he was a guest of King Olaf Tryggvason, who encouraged Leif to convert to Christianity. As a result, Leif also agreed to bring a priest back to Greenland to spread Christianity there.

Upon his return to Greenland, Leif is thought to have met a sailor, Bjarni Herjolfsson, whose ship had veered off course because of a storm. Bjarni arrived in Greenland and reported seeing a green, forested land that he knew was not Greenland because there were no glaciers or fjords. The land he saw is thought to have been the coast of Nova Scotia, making Bjarni the first European to see North America. Leif was an excellent seaman, and he wanted to follow up on Bjarni's story. He knew wood was in short supply in Greenland, and so if they were to find a heavily forested area that was relatively accessible, this would be good for Greenland economically and (of course) profitable for Leif.

With a crew of thirty-five men, Leif sailed in the direction indicated by Bjarni. They soon landed at a flat, stony land he called "Helluland" (possibly Baffin Island or Labrador). He continued south along the coast and stopped in a second location he called "Markland" (thought to be Newfoundland).

How far he actually sailed is in dispute. Some think he traveled as far south as Cape Cod; others feel he disembarked and journeyed inward. The sagas record another stop that Leif referred to as "Vinland" (possibly Nova Scotia), and describe the land as rich, fertile, and forested, "where grapes grow." The men wintered in the new land, and that spring they loaded their ship with some of their newly found bounty and sailed back to Greenland. Leif's success was noted, and he became known as "Leif the Lucky." His father, Erik, died within the year and Leif took over his farm, never traveling again. Other Norsemen seemed to have repeated Leif's journey, but unfriendly Native Americans may have driven them out after only a few years.

Many years passed until the nineteenth century, when a Danish scholar, Carl Christian Rafn, began studying the sagas to examine whether they were actually fact-based. During Rafn's lifetime—and for the next hundred years—no one was able to find any evidence of a Norse settlement anywhere along the North American coastline. But Rafn had gotten the ball rolling. In 1960 a Norwegian historian, Helge Ingstad, and his daughter, Benedicte, came to North America and traveled from Maine to Labrador in a continuing pursuit of this mystery. In Newfoundland, Ingstad heard about a fisherman who had noted some oddly shaped mounds near the town of L'Anse aux Meadows on the northern tip of Newfoundland; the fellow thought they resembled Indian burial mounds. Ingstad thought it was worth investigating the area. In 1961 he returned with his wife, an archaeologist, and they began an exploratory dig at the site. They soon excavated conclusive proof that a Norse settlement from Leif's time had existed there. (Some historians believe that Leif's "Vinland" was actually farther south and has yet to be identified.)

The definitive discovery of a Norse settlement some five hundred years before Columbus brought new facts to the geography and history books about the discovery of North America. In 1964 President Lyndon B. Johnson declared October 9 to be Leif Eriksson Day.

Archaeological evidence indicates that the Norse returned on many occasions over the next few hundred years, but for some reason this did not result in the opening of North America

to newcomers. While the Norse were certainly the first, the credit for opening North America to exploration by Europeans still goes to Christopher Columbus.

One other misunderstanding about Leif Eriksson: History books frequently refer to him as a Viking, but he was not. "Viking" is a term used to refer to pirates that traveled the oceans in the eighth to tenth centuries. Leif was a Christian, and he did no pirating.

Marco Polo (1254–1324 C.E.): Mythmaker Extraordinaire?

If a gathering of Marco Polo scholars were held today, there is little doubt but what fur would fly. Some well-respected scholars believe that Marco Polo never even made it to China, despite having become well known in geography books for his travel experiences there. Others feel that the story of his route is basically true but that some of the experiences he documented were actually the result of conversations he had with other travelers.

So was Marco Polo a great adventurer, as we read in the geography books, or was he less a traveler and more a great "embroiderer of the truth"? Here's what we (sort of) know about Marco Polo.

Marco Polo is thought to have grown up in Venice. His father, Niccolò, and his uncle Maffeo were businessmen in the Middle East, and they returned to Venice after spending more than a dozen years in Constantinople. (Marco may not have met his father until he was a young teenager, as the father was not in Italy for a good number of years.) After returning to Venice, Niccolò and Maffeo traveled from there to what is now Beijing, where the Grand Khan Kublai ruled. Khan wanted missionaries to come to China, so he sent Niccolò and Maffeo back to Italy to meet with Pope Clement IV. Unfortunately, the pope died in 1268, as the two brothers were returning to Italy, so they had little choice but to wait until a new pope was chosen in 1271. They finally had an audience with Pope Gregory; having delivered their message, they were ready to return to Mongolia. The men, accompanied by Marco, left Venice and traveled along the Silk Road through what is now Iran, Iraq, Afghanistan, and China. They are thought to have spent the next seventeen years in Mongolia; some say they carried out diplomatic missions for Khan, while others report that Marco worked as a tax assessor for Khan and was well liked because he was a good storyteller.

After this period in Mongolia, Marco Polo returned to his birthplace and participated in a skirmish that was part of an ongoing battle between Venice and Genoa for commercial dominance of the area. He was captured by the Genoese and imprisoned for a year (1298), when he is thought to have met fellow prisoner Rustichello da Pisa, who was a great romance writer of the day. Polo may have dictated the story of his travels to Rustichello. Eventually a book by Polo was published—*Il Milione*. (Polo's family nickname was Emilione, so the title may have been in reference to that; the book was subtitled *Le divisament dou monde* ["The description of the world"]). For a good many years, Marco Polo's story was considered a relatively accurate report of his time traveling through China; one section even related happenings that concerned a beautiful island kingdom called Cipangu (Japan). By 1375, the places he mentions were beginning to be added to maps. Polo is not thought to have written the book himself, primarily because the original manuscript was written in Old French, a language Polo did not speak.

Over time, readers started noticing descriptions that didn't ring true. One version of the book notes that Mongol royalty hunted with twenty thousand dog handlers and ten thousand falconers—numbers that simply don't make sense considering the population at the time. Various copies and versions showed marked differences, and experts began to look back to identify the most accurate copy. They soon realized that embellishment could have started at the beginning. Polo was supposed to have been a great storyteller, and Rustichello was a novelist. The original version has been lost, and there are now thought to be some 140 different manuscript versions of the text in a dozen different languages and dialects. The book also predated the printing press; each time the book was recopied there were opportunities for changes in the story, creating an immensely

complex and controversial body of material from which to tease out what actually happened.

But the confusion doesn't end there. After Polo's release from prison he returned home to Venice and lived as a wealthy merchant. Then between 1310 and 1320, Polo wrote a new version of his book in Italian. This original text was also lost, but before that happened a Franciscan friar had translated it into Latin. The Latin version was then translated back into Italian, creating even more confusion over the various versions of the story.

So how accurate is Marco Polo's book? Did he really visit the Mongol empire or did he write down the stories of others? And how much of it is embellished? There are well-respected experts with very different opinions. Dr. Frances Wood, a scholar at the British Museum, is among those who believe that Marco Polo never really reached China. She notes that Polo never mentions the practice of Chinese footbinding, calligraphy, or the Great Wall, all of which are the kinds of details that one would expect a traveler from the West to observe.

Yet others have traced the route and believe that in general, Marco Polo's report is accurate. Using a system of measuring distances according to the number of days required to travel from one spot to another, Marco Polo provided a way for others to check the veracity of where he had been. In the nineteenth century, Henry Yule, a Scottish geographer, attempted to verify Polo's route, and Henry deemed that he was among the first to trace a route across Asia—that the story was accurate.

A great number of years later, in the 1990s, the National Geographic sent out writer Mike Edwards to verify the trip. Edwards used what he identified as the most authentic version of the *Travels of Marco Polo,* as the book is commonly known, which was a fourteenth-century copy from the French National Library. Based on his travels and his efforts to document the places described, Edwards believed that Polo made the trip. (His article appears in the *Smithsonian* magazine, July 2008.)

Whether Marco Polo's stories are true or embellished, his writings changed history because they brought a new awareness of the East. Though the Polo family was not the first group of Europeans to reach China overland, their trip—or whatever part of it they made—was the best documented of its time. By exposing China's culture to Europeans, Marco Polo fostered travel between the two areas. His description of the riches of the Far East even inspired Christopher Columbus to try to reach those lands by a western route. (A heavily annotated copy of *Il Milione* was among Columbus's belongings.) Spices (pepper, cinnamon, nutmeg, ginger, cloves) were highly desired by Europeans because they made otherwise bland or spoiled food taste better. Explorers could get rich if they developed faster, better routes to the East, so they were eager to try new approaches. The Italian city-states of Venice and Genoa and the Turkish merchants in Constantinople (present-day Istanbul) controlled the land routes, and their ships dominated the Mediterranean, where they also actively traded with the Arab world.

Polo died in 1324 and a story goes that on his deathbed, he was asked to retract his "fables." He is reported to have said, "I have not told half of what I saw." His legacy was an intriguing description of the world that fostered the interest of others in more fully exploring the East.

Ironically, Marco Polo's route to the East—overland along the Silk Road on a path that had been used since the first millennium B.C.E.—fell into disfavor only about 150 years after Polo's time. By the late 1400s, sea routes were becoming much more popular than the overland routes.

Learning Who Went Where When

The historical record from ancient times is spotty at best. Sometimes information is assumed because of cultural evidence. Many ancient peoples did not have a written language; sometimes there are reports on the explorations of these groups, but they are written by someone who heard tales and wrote them down a couple of hundred years later. Documentation by the Greek historian and geographer Herodotus is valuable, but many of his reports are not eyewitness accounts. Though Herodotus was writing some two hundred years after the Phoenicians sailed the seas, his reports seemed authentic.

For example, Herodotus noted that the sailors described that in sailing around Libya they had the sun on their right hand (meaning they were sailing north and east). Experts look at this as proof that these early voyagers did, indeed, sail around Africa in the seventh century B.C.E.

Translations often prove problematic. The key to Egyptian hieroglyphic translations was not found until the nineteenth century: the Rosetta Stone, a 196 B.C.E. stone that featured a single written passage in three languages—two Egyptian tongues (Demotic and hieroglyphic) and also classical Greek. The stone was found in 1799, but it took scholars almost twenty-five years to develop a translation; it provided a key that permitted the unlocking of many other Egyptian documents.

If written documents are not necessarily contemporary and translations are problematic, historians and scientists are left looking for other clues. Art is always helpful. We know that people traveled the seas during the late Paleolithic period, as there are depictions of Egyptian ships that date to 6000 B.C.E.

The columns and walls of this Egyptian temple are covered in hieroglyphics, an ancient writing system that modern scholars were unable to translate until the nineteenth century. Source: Maison Bonfils, Prints & Photographs Division, Library of Congress, LC-DIG-ppmsca-03923

And people must have traveled western parts of the Silk Road as early as the fourth millennium B.C.E.; there was only one source of lapis lazuli in the ancient world (Badakshan, now northeastern Afghanistan), and lapis lazuli is found as far as away as Mesopotamia and Egypt.

When it comes to deciding when or whether a certain group ever reached a specific destination, the work of the archaeologist has become key. We feel reasonably certain that Leif Eriksson visited North America because archaeological findings verify Norse items from that time period in what is now Canada, verifying the Norse saga that indicates it was Leif who traveled to North America.

Advancing technology has been a great help in better understanding our world. Some technology helps in the "finding" and some helps in the analysis. An example of better "finding" is the fact that in 1999 Robert Ballard, the American explorer who used new technology to locate the *Titanic*, used an underwater robot, deep-water tracking equipment, and a global positioning system to locate two intact Phoenician ships thirty miles off the coast of Israel, which sailed the oceans 2,500 years ago. Another team had been searching for an Israeli sub that had sunk thirty-five years ago, and they noted what seemed to be an ancient shipwreck. This sent Ballard to the area with his deep-sea-searching equipment. The vessels are the first intact Phoenician ships ever found, and they will be very helpful to historians and archaeologists. The vessels' cargoes were hundreds of containers of wine, probably bound for Carthage. Because the ships were so deep—far from sunlight and with so much pressure around them for much collection of sediment—they were perfectly preserved.

Improved methods of study are offered by a study undertaken by Professor Mark McMenamin, a geologist based at Mount Holyoke College. In 1996, he thought of using the enhancement possible through more recent computers to better examine some gold coins that had been found in North America. He presented what he deemed groundbreaking evidence that mariners of ancient Carthage made it to America some time around 350 B.C.E., long before Leif Eriksson and Christopher Columbus. McMenamin undertook studies of gold coins

that were minted in the North African city of Carthage between 350 B.C.E. and 320 B.C.E., and using a computer to enhance the images on the coin, he interpreted a series of designs that he believes represent a map of the ancient world. In addition to reflecting the area around the Mediterranean Sea, McMenamin points out a landmass representing the Americas. If he is accurate, it means that Carthaginian explorers must have sailed to the New World about 1,300 years before Leif Eriksson.

Since scientists understand this, they actually resist uncovering all that they can so that they can save some of the evidence for future scientists working with even better technology. An archaeologist working at the Jamestown settlement noted that they had completed what they had intended to do, and they were in the process of closing up part of the site. "New technology will come along, and we want to preserve some of the materials so that they can be uncovered and studied more completely than we can do today."

These very early explorers set out with little more than "a wing and a prayer," and yet they began to realize that the world was a vast and exciting place. Soon others were to follow.

2

The Age of Discovery Begins with Prince Henry

The geography of the world that is most familiar to Americans today was not identified until the Age of Discovery (the early fifteenth century through the seventeenth century). The stories of these explorers deserve another look. Today scholars have access to more material—more journals have been located and translated, more viewpoints are now represented—and as a result, a good number of these explorers are worth fresh study. A few never made it into the geography books. Others made it in, but their stories were either inaccurate or incomplete.

Starting in the fifteenth century, European national imperialism and economic competition was growing, and competition was particularly fierce between the two countries that dominated the seas at that time—Spain and Portugal. Both countries wanted to claim lands, create colonies, find gold, and develop trade routes to increase their share of the spice trade (spices were popular for use in both cooking and creating medicines). They also believed that it was important to spread Christianity.

Their best means of exploration continued to be via the seas. The countries of Europe had had a safe land passage to China and India (areas that had proven to be rich sources of valued goods such as silk, spices, and opiates), but in 1453 when Constantinople fell to the Ottoman Turks, the land route became dangerous. Islamic powers also made the more easily accessible sea route (south from the Red Sea) very hard to access. As a result of these conditions, the explorers of the time looked south and west with the hope of finding new routes. Fifteenth-century Europeans had no idea of the existence of the North and South American continents, so the idea of traveling west to reach the East made sense to some of the adventurers. Others were intent on sailing south to see if a route around Africa might be possible. The waters just south of the Canary Islands were difficult to navigate, so they had not yet successfully ventured this way.

The captain and his men sailed under less-than-ideal circumstances. The type of sailing ship was relatively new, and their navigational techniques were primitive. (The ships and the navigational methods will be described in chapter 5.) In addition, voyages to new areas rarely had much of a map to rely upon—their guidance usually came verbally from someone who had "almost" reached there. Well-funded groups were fortunate enough to have a supply ship that accompanied them, but even then, for a trip that was to extend for a year or more, the supplies were far from adequate. Staples for the men were generally bread, beer, fish, and salted meat. Slightly better food was generally made available for the captain, who might have had access to a supply of wine and fresh livestock that was kept on board to be turned into food on the trip.

Prince Henry the Navigator (1394–1460)

The people we generally remember as the "discoverers" of our world are the explorers who actually traveled to the lands and claimed them, but as with all ventures, there are earlier behind-the-scenes people who have the ideas, gather the funds, and then create excitement about the idea of what is possible. Prince Henry the Navigator was just this sort of person, and to him the world essentially owes credit for launching the Age of Discovery. Yet it is rare to read about him in any books.

Henry was born in 1394 and was the third son of King Joao I (John) of Portugal. Portugal was just emerging from a period of civil strife and the nobility was impoverished, so John and others began to look abroad for new lands to conquer that might bring both honor and wealth. John was particularly interested in North Africa, an area occupied by Moors right across the sea from Portugal. He and his sons felt that if

Prince Henry the Navigator funded numerous expeditions around Africa and was integral in launching the Age of Discovery. Source: *The Iconographic Encyclopedia of Science, Literature, and Art* (1851)

they could gain a foothold in Africa, they could then explore the coast of Africa as well as the islands just west of that continent. Riches would certainly follow. In 1415, Henry and his brothers were among the soldiers who attacked and took over Ceuta, a port on the North African coast, just opposite the Iberian peninsula at the Strait of Gibraltar. Henry noted the rich results from the Muslim trading with India and interior Africa, bringing home spices, oriental rugs, and gold and silver. This information, coupled with Henry's fascination with navigating the sea and his interest in locating Prester John (see sidebar), made Henry intent on further exploration.

Henry's dreams soon became possible because of two important events in his life. In 1419 his father made him governor of Portugal's southernmost coasts, and in 1420 Henry, who was a highly religious man, was appointed governor of the Order of Christ, a financially wealthy Portuguese successor to the Knights Templar. He held the position for the rest of his life, and because the trips involved converting more people to

Legend Encourages Exploration

Just as the Seven Cities of Gold or the search for the Fountain of Youth would eventually open up exploration in the American continent, stories of a Christian land, headed by a king known as Prester John, encouraged exploration in Asia and Africa as explorers sought out this mythical kingdom.

As early as the twelfth century, a letter describing a mysterious kingdom in the East started circulating around Europe. The letter was thought to have been written by Prester (a form of the word "presbyter" or "priest") John. It was said he was descended from the race of the three wise men. The scepter he used was said to be of solid emerald. This was a time when the Christians in Europe were feeling very threatened by the Muslims, who were overtaking new territory. The thought of a Christian "savior" who headed a powerful kingdom was very intriguing, and many felt he might help the Christians hold their own. Prester John said he ruled a huge Christian kingdom in the East, comprising the "three Indias." He described his kingdom as a crime-free kingdom where "honey flows . . . and milk abounds."

Over the following few centuries, various forms of this letter kept reappearing, possibly as many as a hundred different versions. The letter was generally addressed to Emanuel I, the Byzantine emperor of Rome, but some versions were addressed to the pope or the king of France. Countless explorers ventured out to locate Prester John and his kingdom. The legend reported that its rivers were filled with gold and it was home to the Fountain of Youth, so people were very eager to find this land.

By the fourteenth century, explorers had proven that Prester John's kingdom was not in Asia, so they began to focus on Abyssinia (present-day Ethiopia). Throughout the fifteenth century, Portugal sent expeditions to look for Prester John. Cartographers continued to keep the kingdom on maps through the seventeenth century. Over time, the letters became more detailed and more elaborate.

Experts disagree on who wrote the original letter and how it grew into a legend. Many think the basis for Prester John's kingdom was the empire of Genghis Khan. Others feel it was simply fantasy. Though scholars have yet to identify how or where the first letter actually originated, Prester John deserves a place in geography books for his profound effect on the geographical knowledge of Europe, because the story stimulated so much interest in foreign lands.

Christianity, he could use the funds to launch the explorations he deemed worthy.

From 1419 until his death in 1460, Henry sent expedition after expedition down the west coast of Africa to outflank the Muslim hold on trade routes and establish colonies. His influence—and his willingness to finance the trips—led to the rounding of Africa by Portugal and the establishment of sea routes to the Indies. This was a major accomplishment because Africa had defeated many fine sailors. The Spanish had asserted for generations that it was not possible to go beyond the Canary Islands because of reefs and currents that tossed boats around. For that purpose alone, Henry commissioned fifteen expeditions between 1424 and 1434, before a vessel finally made it to Cape Bojador, south of the area that was formerly impassable. Eventually the exploration became very beneficial to Portugal and Henry was enriched by it.

Henry became a legendary character, and it is widely reported that he established a navigation school that greatly influenced the world. But some of the reports on Henry's accomplishments were written by contemporaries who wanted to curry favor with the prince. Henry encouraged exploration, and he hired cartographers to map new lands, but new scholarship reveals no evidence of any formalized nautical school from that time.

Twenty-seven years after Henry's death, Bartholomeu Dias finally traveled beyond the equator and found his way around the Cape of Good Hope in 1487. This opened a path for Vasco da Gama to follow. (See "Vasco da Gama" later in the chapter.)

Christopher Columbus (Cristoforo Colombo) (1451–1506)

For an explorer who has a federal holiday named for him, Christopher Columbus is being knocked off his pedestal by blows from all sides. Despite what schoolchildren have been taught, Columbus was not the only person to recognize that the earth was not flat, nor was he the first to discover the "new world."

In the late 1980s, scholars undertook new research to prepare for Columbus's quincentennial celebration in 1992, and they found that most learned men of Columbus's day knew the earth was not flat. So sailing west to reach something in the East was not as revolutionary a thought as we were originally taught. (Actually, the spherical view of the world dated to the Greeks, who realized this as early as the fourth century B.C.E.)

In more recent times, we have also learned that Columbus was not the first to discover the New World (see Leif Eriksson, chapter 1), and he actually never reached North America at all. Nor did he realize he had reached a "new world." (The reasoning behind the state of Connecticut naming a highway after him is truly baffling.) Columbus died still thinking he had reached islands off India, which is why this area is also known as the West Indies.

Christopher Columbus. Source: Etching by F. Focillon after painting by Bartholomeo de Suardo, Prints & Photographs Division, Library of Congress, LC-USZ62-91021

The most recent flurry of news surrounding Columbus has concerned exactly who the Great Navigator actually was, and this question is being asked because of the new possibilities being offered by DNA analysis. In 2004 a Spanish geneticist, Dr. Jose A. Lorente, was able to extract a small amount of genetic material from Columbus's bones, and Dr. Lorente has been beset by people who want him to investigate all types of questions about the explorer's background.

What has been determined thus far is that there was a Cristoforo Colombo from Genoa, but it is becoming less clear whether this is the "correct" Columbus. Scientists have studied his handwriting and evaluated the letters he wrote and have noted that he wrote in Castilian (a kingdom that was part of Spain), kept books that were written in Catalan (a language that developed in the Pyrenees mountains in Spain), and married a Portuguese noblewoman. He decorated his letters with a Hebrew cartouche (an oval decoration as on maps). All these factors lead the experts to a variety of conclusions. Some think Colombo may have been born out of wedlock to a Portuguese prince (hence the eventual good marriage), or he may have been a rebel in the kingdom of Catalonia, or he may even have been a Jew whose parents converted to escape the Spanish Inquisition. At this point, no definitive study has been conducted because it was not possible to extract more than a small trace of DNA. Dr. Lorente says it is important to wait until there are more advances in science or more evidence from other arenas that permit asking as many questions as possible once the DNA is examined again.

Though the story of Columbus proved to be largely a myth, scholars today still give him his due in one particular area. He opened the Americas to exploration by demonstrating that if a ship sailed west, the captain and crew would reach a rich land. He opened a pathway from the Old World to the New and forever changed life on both sides of the Atlantic.

While this proved excellent for trading and empire building, we now also know that Columbus's accomplishments came at the expense of the native people. As the story has unfolded more completely, we have learned that Columbus and his men often treated the native Taino people barbarically,

enslaving many to work in mines so that Columbus could take home riches. So while Columbus still deserves a place in the history and geography books, perhaps his credentials are being more carefully identified. Here is what scholars currently think happened in Columbus's life.

Columbus's Career

Columbus's career as a seaman began when he and his brother Bartholomew were employed in Lisbon as chart makers. Columbus preferred seafaring, and whenever possible he went out with the merchant marine and traded goods along the coast of West Africa. In his travels, he grew familiar with the Atlantic wind systems, which eventually permitted him to benefit from the flow of the trade winds to navigate to what we now think of as the West Indies.

Columbus believed that the ocean to the west of Europe was completely open, and he reasoned that if the world was spherical, he could reach the Far East by sailing west. (No one knew of the existence of the continents of North or South America.) In 1484 Columbus approached the Portuguese king to ask him to support an Atlantic crossing in search of a new way to India, but Columbus's request was rejected. Two years later, he traveled to Spain and asked King Ferdinand and Queen Isabella for their support in locating a new route to China. Columbus promised to return with gold, spices, and silk and to spread Christianity as he went. The king and queen knew what this discovery would mean to Spain if they "owned" a new and faster way to reach the lucrative gold and spice market of the Indies, but they could not commit to the trip at that time. To keep him from taking his ideas elsewhere, Ferdinand and Isabella offered Columbus an annuity. Six years later (1492), they finally committed to funding his trip. Ferdinand and Isabella told him if he was successful, he could become governor of all the discovered lands, and they also promised him the honorary title "Admiral of the Ocean Seas."

On August 3, 1492, Columbus departed from Palos, Spain, with 104 men on three ships. Columbus relied on his sailing experience, observation, guesswork, and intuition to judge his ships' positions and the direction they needed to sail. (Also see chapter 5 for more information on their navigational equipment.) They averaged about 150 miles per day, but Columbus had misjudged the earth's circumference and thought the journey would be shorter than it was. They did not see land until October 12, 1492—a long period to be afloat in the middle of the ocean. Columbus declared they had reached India, but it was actually one of the islands in the Bahamas. They spent the next few weeks exploring parts of Cuba and the northern coast of Hispaniola—thinking, of course, that they were investigating new islands off India.

On this visit, Columbus and his men were greeted warmly by the natives, and Columbus referred to them as "Indians" since that was where he thought he was. Before sailing back to Spain, he stationed thirty-nine Spanish sailors on Hispaniola in what is now Haiti and captured an estimated ten to twenty-five native people to take back with him to present to the king and queen (only seven or eight of the natives survived the crossing).

Due to the overwhelming success of the 1492 venture, his second trip was better funded, and he had a full contingent of seventeen ships with him that time. When they reached Hispaniola, he learned that the Spaniards he had left behind had been killed by the natives. As the story has become clearer, it is thought that the Spanish sailors mistreated the natives, and the Tainos (as they are known) retaliated by killing the men.

On Columbus's third voyage, he successfully reached the mainland of South America. But when he returned to Hispaniola he was met by rebellion from both the native people and the Spaniards who had recently arrived with him. They accused him of tyranny, and the men resolved that he would be sent back to Spain in chains. Upon his return, the king and queen were distressed but agreed to meet with him. After hearing Columbus's side of the story, they restored his wealth and agreed to fund a fourth voyage (1502–1504) to look for a strait to the Indian Ocean. They decided he should no longer be given the power to govern the new lands, and he was told not to land in Hispaniola for fear he would stir up trouble again.

On this trip, he made it to the mainland of Central America for the first time, but Columbus noted in his journal (December 5, 1502) that it was a particularly challenging trip:

> For nine days I was as one lost, without hope of life. Eyes never beheld the sea so angry, so high, so covered with foam. The wind not only prevented our progress, but offered no opportunity to run behind any headland for shelter; hence we were forced to keep out in this bloody ocean, seething like a pot on a hot fire. Never did the sky look more terrible; for one whole day and night it blazed like a furnace and the lightning broke with such violence that each time I wondered if it had carried off my spars and sails; . . . [the rain was a deluge]. . . . The men were so worn out that they longed for death to end their dreadful suffering. (From a biography of Columbus, *Admiral of the Ocean Sea,* written by Samuel Eliot Morison in 1942.)

Marooned for a year on the island of Jamaica, Columbus again encountered trouble with the local people by overstaying his welcome. Initially, they had been willing to help the Spaniards, but as a full year dragged on, they tired of the men's dependence and cut off the supplies they were providing. Columbus needed his crew to be fed and came up with a plan. He knew from his almanac that a lunar eclipse was coming, so he told the native people that he would "take away the light of the moon" if they displeased him. On the night of February 29, 1504, the eclipse of the moon began. Convinced that Columbus was holding good to his threat, the Indians became alarmed and agreed to continue to provide for them.

After this trip, Columbus went home a disappointed man; he died in 1506 not having found the spice source and great wealth that he had expected to find in "India." Yet he was still convinced he had set foot in Asia rather than the New World.

Juan Ponce de León (1460–1521): Pathmaker

Ponce de León, a Spanish conquistador, has gone down in the history books for his quixotic search for the Fountain of Youth, but he actually deserves more prominent recognition for establishing a vital stepping-stone for Spain's entry to the

Did Columbus and His Men Bring Syphilis to Europe?

It has long been suggested that the syphilis that spread across Europe during the 1500s and 1600s was brought back from the New World by Columbus and his men. Many of the crew members who served on this voyage later joined the army of King Charles VIII in his invasion of Italy in 1495. Around that time, the disease began to spread across Europe, causing as many as five million deaths.

Based on newly possible genetic studies, researchers at Emory University have used phylogenetics (the study of evolutionary relatedness between organisms) to examine various strains of the disease that are known to be from different parts of the world. They have been able to link the venereal syphilis strain—the one that spread among Europeans—to the South American strain that causes yaws. This would support the hypothesis that syphilis originated in the New World.

Not all researchers agree with this conclusion, but they all do agree that more study is necessary. One of the coauthors of the study, Emory skeletal biologist George Armelagos, says: "Syphilis was a major killer in Europe during the Renaissance. Understanding its evolution is important not just for biology, but for understanding social and political history. It could be argued that syphilis is one of the important early examples of globalization and disease."

New World. His early voyages to the coast of Florida paved the way for the Spanish to build a fort at St. Augustine, a remarkable fifty-five years before the Pilgrims arrived at Plymouth Rock and a full forty-two years before the founding of the English colony at Jamestown.

Ponce de León first came to the New World with Christopher Columbus on Columbus's second voyage to the area. It is believed they first landed on the Turks and Caicos islands, but they soon sailed on to Hispaniola (the island now shared by Haiti and the Dominican Republic). When Columbus was preparing to return to Spain, Ponce de León opted to stay on Hispaniola. He began to hear stories of the wealth of a land called "Borinquen" (Puerto Rico), and he wanted to explore the area. In 1509 Ponce de León made his way to Puerto Rico with his sailors and founded the first Spanish settlement there,

Juan Ponce de León. Source: John Ledyard Denison, Prints & Photographs Division, Library of Congress, LC-USZ62-3106

Caparra. He was soon named governor by the Spanish crown. Ponce de León was intent on developing mines and bringing out wealth, and he attempted to use the local people, the Tainos, to construct fortifications and conduct the mining. The community did not do well. Ponce de León coerced the natives into working for him, and many people died from the brutality. In addition, the European sailors introduced diseases to which the local people had no immunity, and this caused even more deaths.

Christopher Columbus died in 1506. His son Diego had observed what was happening in Puerto Rico and took issue with Ponce de León's governorship. He took his complaint to the top court in Madrid, claiming that his father had been deemed "governor" of all lands and that by inheritance, this title and responsibility should have passed to Diego. The Spanish government removed Ponce de León but felt he could still be of value to the country, so the king encouraged him to continue to explore and claim new lands. Ponce de León learned from the Indians of an island called "Bimini" (in the Bahamas)

on which there was a water that could rejuvenate those who drank from it. So Ponce de León set out to find this vitality-restoring water. This took him north of Cuba, and he traveled as far as the northeast coast of Florida, arriving there in March of 1513. He claimed the land for Spain and called it "La Florida," perhaps after the vegetation in bloom there or perhaps for the fact that he arrived during "Pascua Florida" (Palm Sunday). Later he sailed along the southern coast, passing the Florida Keys, and up the west coast as well.

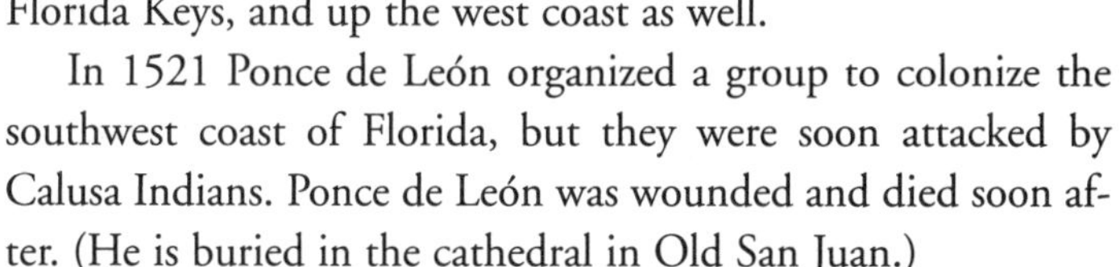

In 1521 Ponce de León organized a group to colonize the southwest coast of Florida, but they were soon attacked by Calusa Indians. Ponce de León was wounded and died soon after. (He is buried in the cathedral in Old San Juan.)

While Ponce de León was not the first European to reach Florida (John Cabot likely preceded him), he discovered the Gulf Stream oceanic current, which was to become vital to many who were sailing from Europe to the New World and back again. During his travels, Ponce de León also accomplished great things for Spain by making North America accessible to the Spanish. His accomplishments became the proverbial

So What Happened to Florida?

The French established a foothold in Florida in 1564 and built a fort on the St. Johns River (just south of St. Augustine), and they began threatening the Spanish ships that passed by. In 1565 the Spanish king took a definitive step to claim what he felt was his, and he sent one of Spain's most experienced admirals, Pedro Menéndez de Avilés, to become governor of Florida. Menéndez de Avilés quickly took control of the area and soon defeated the French on land and at sea. The Spanish saw that the British were making inroads in the area, so they built a fort at St. Augustine. Spain was able to maintain control of Florida until after 1763, when Spain ceded Florida to England in order to regain the capital of Cuba. The area was then ruled by the British for about twenty years until the Revolutionary War, when things changed again. Under the Treaty of Paris (1783), Florida was returned to Spanish rule for a period of thirty-seven years, and when Spain sold Florida to the United States, the Spanish departed for the last time. A transfer-of-ownership ceremony was held in 1821, and Florida became a state in 1845.

"one step forward" before "two steps back." Between 1513 and 1563 the government of Spain launched six expeditions attempting to settle Florida, but all failed.

Vasco da Gama (ca. 1460–1524): New Route to India

Vasco da Gama was a Portuguese navigator who made three trips to Asia via the Cape of Good Hope, ushering in a new era of world history with having opened up a sea route from western Europe to the East. Today he is celebrated by the Portuguese but vilified by those in India who feel he brought about the beginning of repression and colonialism in India. What's the real truth?

As mentioned previously, Portugal believed the best way to sail to India was to navigate south, around the Cape of Good Hope. In 1487, when Bartholomeu Dias (1450–1500) sailed around the southernmost tip of Africa, he was able to verify that the coastline turned and extended to the northeast; this seemed very promising. Ongoing concurrent land exploration gave the Portuguese continued hope that India was accessible from the Atlantic Ocean. Muslim domination of the trade routes was weakening, so the Portuguese prepared to open their own sea route to India. Estêvão da Gama (ca. 1430–1497) was chosen to lead the expedition, but he died before the trip got under way; his son, Vasco, received the commission in his stead. King Manuel specified that da Gama should secure for Portugal access to the great markets of Asia and seek out Christian kingdoms in the East. With four small ships, 171 men, and food reserves for an estimated three years, da Gama took off from the western tip of Europe, from Lisbon, and navigated down and around the coast of Africa. The Portuguese maintained a custom that was helpful to other travelers and eventually to historians as well. As they explored new territory, they built *padrões*, stone pillars that featured insignia of their current ruler (the royal arms of King John II) as well as a Christian cross. These, plus contemporary reports, provide us with information about da Gama's first journey to India.

Da Gama got off to a good start but he soon got knocked off course. He was familiar with the wind patterns of the Atlantic so he put his ships on a south by southeast course before making a wide sweep westward to reach the currents and winds he would need to round the Cape of Good Hope. Unfortunately, his ships got caught up in the currents of the Gulf of Guinea, and his ships traveled for ninety-three days out of sight of land, covering six thousand kilometers, and they barely reached the cape. (The distance traveled was three times the distance traveled by Columbus during his first voyage.)

Da Gama's progress on the southeastern coast of Africa was tedious. Many of the men were suffering from scurvy (a terrible illness involving bleeding gums and excessive muscle weakness that results from a lack of vitamin C), and they met with many hostile native people. When they reached what is now Mozambique, da Gama had little choice but to stop, get the ships repaired, and let his men rest. To quell the hostility of the Muslim residents of the island, da Gama led the residents to believe that the sailors were Muslim, too. In what is now Kenya, they met up with a navigator who knew the route to Calicut on the southwest coast of India. The fellow led them on a twenty-three-day run across the Indian Ocean, where da Gama was victoriously able to erect a *padrão* to prove he had reached India. On May 20, 1498, they had finally found a southern sea route to the Asian markets. Partly because many of the Muslim merchants were hostile to the Portuguese coming into the area and partly because he had no valuable goods to trade, da Gama was unable to conclude a treaty with the Hindu ruler of the area.

In some frustration, da Gama was ready to go home at the end of August. He ignored the local people who tried to explain about the monsoon wind patterns and prepared to depart, taking with him five or six Hindus so that the Portuguese might see and learn about them. Instead of the twenty-three-day transit time of their first crossing, the monsoon winds pushed the vessels in the other direction, and the return trip across that stretch of ocean took a horrific 132 days. During the trip approximately half the crew died and many of the

others were affected by scurvy. Only two of the ships finally made it home to Portugal, seven or eight months later.

Vasco da Gama was greeted with great honor when he returned to Portugal. He was made an admiral and also awarded the perpetual title of *don* (lord). This was the beginning of the European nations controlling the seas, an era that lasted until the end of the nineteenth century, when Japan and the United States became major naval powers.

Five hundred years later, it's a different story. In 1998 Portugal held major celebrations to commemorate the occasion, but India responded with flying black flags and burning da Gama in effigy in some communities, noting that da Gama marked the beginning of colonialism and repression in India. Da Gama supporters say that there is no direct line between da Gama and colonialism in India, and that da Gama's contribution was the opening of the trade routes that changed the nature of world trade and travel.

As much as anyone after Henry the Navigator, da Gama was responsible for Portugal's success as an early sea power. Portugal realized that securing outposts in eastern Africa was vital to maintaining trade routes. The spice trade was a major asset to Portugal, and Portugal also began to benefit from the settlements it then knew of in eastern Africa. As for India, if da Gama had not arrived, someone else would have. India was a huge prize on everyone's "wish list," so Europe was coming whether India wanted it or not.

The lure of the spice trade continued to entice. In the next chapter, you'll meet still more explorers who were learning about the world as they pursued new ways to sail to the East.

3

The Age of Discovery Continues

The Cabots, Vespucci, and Magellan

As word of the discoveries that were being made spread from country to country, Britain stepped forward to participate. Though its fleets were not yet as strong as Spain's and Portugal's, Britain didn't want to miss out on what seemed like important discoveries. The explorers themselves were opportunistic. Patriotism was not a factor in their thinking, as their intense drive to explore sent them from country to country in search of funding for the opportunity to explore.

John Cabot (Giovanni Caboto) (ca. 1450–ca. 1499): An Important "First"

Like other explorers sailing west, John Cabot expected to reach the East. So when Cabot landed in the vicinity of Newfoundland, he assumed he had reached a northeastern island near Asia. Since he and his men found no inhabitants in the area, Cabot decided it was an as-yet-undiscovered island, and he named the area "new found land." He went back to England and announced that he had sailed west and found a new route to Asia. Of course, we now know that Cabot was the first European explorer to set foot in North America since Leif Eriksson's journey in about the year 1000. So how did John Cabot become an explorer who had such an important "first," and why is no federal holiday set aside for him as has been done for Columbus?

John Cabot's father was an Italian merchant who traded in spices in the eastern Mediterranean. John followed his father into the business, and he came to appreciate the great value of the goods (spices, silks, precious stones) being brought in from

Asia. Since he loved the sea, he contemplated finding a new route to Asia, a discovery that would mean great things for the family business.

In 1490 Cabot began looking for funding for an exploratory expedition, but the two major governments that dominated the seas—Spain and Portugal—both turned him down. Both countries were working on their own plans; Spain was financing Columbus's expedition, and Portugal was seeking a new route to Asia by going around the Cape of Good Hope. Cabot decided to approach England; he chose to settle in Bristol, in the northwestern part of the country. He calculated that the longitudinal distances were narrower as one moved north, so if he took a northerly route then his trip would be shorter and quicker. The Bristol merchants were delighted with the business opportunities this type of accomplishment would present for them, so they backed the plan wholeheartedly. Cabot soon gained funding from King Henry VII and set out in 1496, but the trip had to be aborted because of problems with the crew, a shortage of food, and bad weather.

The following year Cabot mounted a second attempt using a smaller ship, the *Matthew*, with only eighteen crewmen. After a stop in Ireland, he sailed across the ocean to what was probably Newfoundland (both Canada and the United Kingdom have deemed Newfoundland as Cabot's official landing spot), but he may have actually landed in Labrador or in Maine or on Cape Breton Island just off Nova Scotia. (All documentation of his travels comes from contemporary reports of his time, but none of those who have written about the trip were among the eighteen who were actually with Cabot.) Regardless of his exact landing spot, Cabot was the first European to land in North America—or either American continent—since Leif Eriksson had landed almost five hundred years previously.

When he returned, the English greeted Cabot as a hero. He was made an admiral and granted a pension for life as a result of his success. A year later Cabot left on his next voyage, traveling west again but with five ships this time. Bad weather forced them to stop at an Irish port so that one of the ships could be repaired after suffering storm damage. Cabot and the others sailed out again, but Cabot and those on his ship were never heard from again. They were presumed to have been lost at sea, though reports indicate that not all of his vessels were lost; some may have returned to Britain eventually but there was no sign of Cabot again.

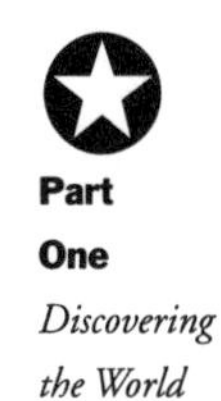

Cabot's achievement was considerable, but our federal holiday is "Columbus Day," not "John Cabot Day," and there are a couple of factors contributing to this. The first certainly has to do with Cabot's untimely death. Because he did not live long enough to make more trips, he was not around to "bang the drum" for himself and what had been accomplished. He was also being backed by a lesser sea power. Britain did not yet

Although most likely John Cabot was not accompanied by his son when he arrived in North America, this artwork shows Sebastian at his side.
Source: Ballou's Pictorial, Prints & Photographs Division, Library of Congress, LC-USZ62-3029

rule the waves, and as a result, a more powerful country like Spain was better able to capitalize on Columbus's achievements.

Sebastian Cabot (Sebastiano Caboto) (1484–1557): The Family Work Continues

For a long time there has been a blurring of the accomplishments of John Cabot and Sebastian Cabot, one of John's three sons. Some of this confusion may have been perpetuated by Sebastian himself, since his accomplishments were considerable but ultimately ended in the failure to reach his goal of locating a northwest passage.

Sebastian was an accomplished mapmaker and navigator in his own right, but he was probably not with his father when his father reached Newfoundland. This expedition took place in 1497 and the ship carried only eighteen men. Sebastian would have been so young at this time that it is unlikely that his father would have allotted space to him. They needed strength and experience for such an arduous journey into the unknown.

After John Cabot disappeared, Sebastian sought and received backing from Henry VII to sail in search of this dreamed-of route to Asia. Sebastian sailed for North America and explored the coast of North America; he is thought to have made it as far as the Hudson Strait (this would have led into Hudson Bay). He arrived back in England to learn that Henry VII had died, and Henry VIII was not interested in supporting this field of exploration. Sebastian moved to Spain, where the Spanish monarch funded him to look for an easy and safe route east.

In 1526 he set out with four ships and spent four years sailing along the east coast of South America. In 1530 he returned to Spain in disgrace because of his failure to find the new route he sought and was not funded for any other journeys.

Amerigo Vespucci (1454–1512): He Recognized the New Land

Columbus may have been the person who first "discovered" the New World during the Age of Discovery, but Italian navigator Amerigo Vespucci deserves credit for actually realizing it was a "new world." The debate as to whether "the Americas" should have been named after him will continue as long as there are people to discuss it. Other confusion also swirls around Vespucci. Until the 1930s, it was accepted that he made four voyages, but since that time scholars have discovered that only two trips can be verified. Here's what else is known based on recent research.

As a boy Vespucci was educated by his uncle. In 1479, he was introduced to the wealthy and powerful Medici family, who soon hired him to work for them. In 1491 Vespucci was transferred to Seville to work in the Medicis' ship-outfitting business, and was eventually placed in charge of it. Among his undertakings was outfitting Christopher Columbus's ships for his second and third trips.

Though Vespucci left letters that documented his travels, his reports are debated. Between variations in translations and in copies, the trail to the source of the letters becomes unclear, and scholars are now not even sure whether Vespucci actually wrote the letters. One letter exists only in translation and specifies four voyages. Three private letters from Vespucci to one of the Medicis also exist. In these letters, Vespucci refers to only two voyages. All agree that Vespucci made at least two trips; whether or not he made a third and fourth trip is hotly debated.

Seven years after Columbus first landed in the West Indies, Vespucci served as navigator for an expedition of Alonso de Ojeda in 1499–1500. Though they landed on the mainland of South America, they assumed they had reached Asia; they explored the northern coast of South America to well beyond the mouth of the Amazon, naming some of the areas, such as "Gulf of the Ganges," based on their supposed location.

When they returned to Spain, Vespucci was turned down by Spain in his request to make another expedition, so he turned to the Portuguese government for funding. He left Lisbon on May 13, 1501. During this trip Vespucci realized he had discovered a totally new continent and set out to verify this by sailing along the coast of South America down to within four hundred miles of Tierra del Fuego (the tip of

Amerigo Vespucci. Source: Prints & Photographs Division, Library of Congress, LC-USZ62-63115

South America). His return route is unknown, but on July 22, 1502, he came back to Lisbon with his report.

This trip of 1501–1502 is of fundamental importance to the history of geographic discovery, for after this voyage he wrote to de Medici that it was not Asia and must be a "new world," a fourth continent after Europe, Asia, and Africa. When he came home, he published an account of the trip between 1502 and 1504, and he—and this information—became widely known.

Vespucci returned to Seville in 1505, received Spanish citizenship, and three years later was appointed "pilot major" for the Casa de Contratación. He was responsible for teaching navigation to those who were sailing for Spain and for drawing maps of newly discovered lands and updating them as needed. Among his accomplishments were great improvements in navigational techniques. He was also skilled at keeping track of distances, and as he traveled he began to understand the earth's circumference, predicting it to within fifty miles.

While verifying all of the information about Vespucci is difficult, most historians feel comfortable with the belief that Vespucci reached the mainland (the South American mainland) a few weeks before Cabot and about fourteen months before Columbus.

So who thought of calling the two new continents "America"? The mapmaker. When Martin Waldseemüller reprinted Vespucci's *Quattuor Americi navigationes* ("Four Voyages of Amerigo") in 1507, he added his own pamphlet about cosmology as well as some maps he had created of Vespucci's travels. He wrote of the new landmass that Vespucci had explored, calling it "America" after Amerigo. This was the beginning of calling the continent America—eventually what would become known of as South America.

Ferdinand Magellan (ca. 1480–1521): Around the World (Almost)

For years schoolchildren have been taught that Ferdinand Magellan was the first to circumnavigate the globe, but it turns out that he didn't actually make it. In 1519 Magellan set out with a total of 237 men on five relatively small ships. Two years later, only one ship and eighteen of those men actually completed the trip; Magellan was not one of them. They had navigated through many rough times, when Magellan became involved in battle with natives in the Philippines and was killed. (We are reminded how very difficult it was to be an explorer departing for parts unknown.)

Though Magellan did not live to complete the trip, he participated in many firsts. He was the first to conceive of a voyage

What Did They See?

As this letter shows, the sight of a new land populated by different-seeming people presented the explorers with much to consider:

> Letter of Amerigo Vespucci
> To Pier Soderini, Gonfalonier of the Republic of Florence

Vespucci noted that the letter was written when their ship was "distantly westward from the isles of Canary . . . about a thousand leagues beyond the inhabited region."

> [W]e anchored with our ships a league and a half from land; and we put out our boats freighted with men and arms: we made towards the land, and before we reached it, had sight of a great number of people who were going along the shore: by which we were much rejoiced: and we observed that they were a naked race: they shewed themselves to stand in fear of us: I believe (it was) because they saw us clothed and of other appearance (than their own): they all withdrew to a hill, and for whatsoever signals we made to them of peace and of friendliness, they would not come to parley with us: so that, as the night was now coming on, and as the ships were anchored in a dangerous place, being on a rough and shelterless coast, we decided to remove from there the next day, and to go in search of some harbour or bay, where we might place our ships in safety: and we sailed with the maestrale wind, thus running along the coast with the land ever in sight, continually in our course observing people along the shore: till after having navigated for two days, we found a place sufficiently secure for the ships, and anchored half a league from land, on which we saw a very great number of people: and this same day we put to land with the boats, and sprang on shore full 40 men in good trim: and still the land's people appeared shy of converse with us, and we were unable to encourage them so much as to make them come to speak with us: and this day we laboured so greatly in giving them of our wares, such as rattles and mirrors, beads, spalline, and other trifles, that some of them took confidence and came to discourse with us: and after having made good friends with them, the night coming on, we took our leave of them and returned to the ships: and the next day when the dawn appeared we saw that there were infinite numbers of people upon the beach, and they had their women and children with them: we went, ashore, and found that they were all laden with their worldly goods.

that permitted circumnavigation of the globe, and he was the first to navigate the southern strait in South America, which connected the Atlantic and the Pacific oceans—a very notable accomplishment. He was also the first to cross the Pacific from east to west, and he saw and documented animals and birds for the first time. But among his most significant contributions may have been ship discipline. Though he was not well liked, his onboard operation ran with exacting precision, and as a result, when one of the ships finally struggled back to the Spanish port, the men knew with great certainty that they had kept careful time logs. As a result of the discrepancy between the Spanish calendar and the shipboard calendar, Magellan's men were the first to realize the need for an International Date Line to create a way to adjust for the time. (Much of what we know about Magellan's voyage came from the diary of a crew member, Antonio Pigafetta, who faithfully recorded where the ships sailed and what happened along the way; fortunately he was among the few who completed the journey.)

Magellan's family was Portuguese nobility, but like other navigators of the time, Magellan sought funding from any country that would sponsor him. Spain and Portugal still reigned supreme at this time, and because they were the superpowers of the day, they decided they could divide the world in two. In 1494 the two countries agreed that Spain would be "given" any discovered lands to the west including the Americas (though Brazil was granted to Portugal); Portugal was to get everything to the east, including Africa and India. The governments had not yet determined where the line should be drawn in Asia, so as a result, the race to the east continued at a great pace.

Vasco da Gama's foray around the Cape of Good Hope provided one route, but the Portuguese found themselves battling Muslims for a piece of the spice trade, so both countries continued to look for a route that wasn't dominated by others. When Magellan proposed his plans to circumnavigate the globe, he was turned down by Portugal but Spain approved the trip in 1518, with King Charles providing him five rather small ships (the *Trinidad*, the *San Antonio*, the *Concepcion*, the *Victoria*, and the *Santiago*).

Ferdinand Magellan. Source: Prints & Photographs Division, Library of Congress, LC-USZ62-30424

Magellan's voyage began with a stop at the Canary Islands, followed by a southwest journey toward Brazil (but not landing there since it was Portuguese territory and Magellan was sailing under the Spanish flag). His ships were manned with Spanish captains and crews who were resentful of Magellan, who was Portuguese and not well liked by most people anyway. They mutinied almost immediately; Magellan quelled this first uprising and they continued on, sailing off the coast of present-day Rio de Janeiro. The weather became so bad that they stopped in Patagonia (Argentina) for the winter and the crew again rebelled. A good number of crew members were either imprisoned, marooned, or executed according to Magellan's orders.

During the summer months, Magellan continued his search for a passage to get to the other side of South America. He finally found the strait that now bears his name, but it was not "smooth sailing." It took his fleet thirty-eight days to pass through, and the captain of the *San Antonio* became disheartened and quit and went back to Spain. In late November Magellan with three of his ships sailed through to what he called *Mar Pacifico.* Magellan estimated that the journey from that point across the ocean to the Spice Islands would be a short one and thought they would cross in two to three days. It actually took four months, and they spent ninety-nine days in the Pacific not seeing any land.

By this time the conditions for the crew were abysmal. Food and water were depleted and because they had no fresh fruit or vegetables, many of the men became too weak from scurvy to help sail the ships. In April they landed at the Philippines but unfortunately, Magellan got involved in a tribal battle that was taking place among the natives, and on April 27, 1521, Magellan was killed.

Juan de Elcano took command of the remaining three ships and 115 survivors, ordering one ship burned because there were too few sailors to crew it. They journeyed on to the Moluccas (the Spice Islands), reaching them in November, where they took time to pick up supplies and items to trade. De Elcano decided that in order to ensure that at least one ship made it back home, he should send the two remaining ships back via different routes. The *Victoria* continued west while the *Trinidad* traveled east across the Pacific. The *Trinidad* was seized by the Portuguese and most of the crew were killed. The *Victoria* sailed across trade routes in the Indian Ocean, rounded the Cape of Good Hope (September 6, 1552), and almost three years from the day the voyage first began, the *Victoria* with eighteen crewmen (including diarist Pigafetta), arrived back in Spain. It was the first vessel to circumnavigate the globe.

Though Magellan and his men had kept scrupulous records of their days away from home through maintaining the ship's log, when they arrived home, they discovered that their calendar was a day behind the Spanish calendar. They did not have

clocks accurate enough to observe the very slight lengthening of each day while they were on their journey (traveling east, the earth's rotation was one less than if they had stayed in Spain). This discovery was greeted with great excitement and a special delegation was sent to the pope to explain what happened and how it occurred. The eventual resolution of the discovery was the creation of the International Date Line.

For the next fifty years, no one was able to follow the navigation path of Magellan until Sir Francis Drake in 1577. Drake (ca. 1540–1596) was a British admiral who successfully circumnavigated the globe and was the most famous seaman of Elizabethan times. The British revered him for the territory he claimed for them; the Spanish reviled him because he plundered many of their settlements, including several in the West Indies. Having been raised by a friend of his family who was both merchant and pirate, Drake followed that family's tradition and was as likely to plunder as he was to discover.

With these explorers, the first wave of the Age of Discovery was under way. Using intuition and relying on basic equipment and a big dose of bravery, these men sailed across the ocean and began to explore new lands that they had never dreamed of. But there were more adventures to come as the Age of Discovery continued, and you'll read about the next wave of explorers in chapter 4.

4

More Men and More Countries Exploring the World

As the world began to open up a bit, the opportunity arose to explore in more detail. There was still one driving force—a search for better routes to the East—and some of this next group of explorers continued on that quest. Others made major discoveries such as the Hudson River, Hudson Bay, and the Mississippi River, all while continuing to look for ways to the East Indies. Because land travel was still so difficult, few paused to think that if they were to explore more fully in the Americas they might find resources that were as valuable as—or more valuable than—the spices they sought so avidly.

Henry Hudson (1570–1611): Unjust End?

Henry Hudson (not "Hendrick," as he is sometimes referred to) is strongly associated with exploring because of the prominent river and bay that are named for him. What isn't well known about Hudson has to do with the "ultimate reward" of being an explorer. Despite his many accomplishments, Hudson and his son were victims of the mutinous uprising of the crew. They were set adrift in Hudson Bay in 1611 by Hudson's own men and were never heard from again.

Little is known about Hudson's boyhood, but it is thought that he started sailing at a very young age. Some say he was a cabin boy at sixteen, working his way up to captain. Because of his particular knowledge of Arctic geography, it is suspected that he may have been exposed to the ideas of some of the early navigators who had made attempts at finding a northwest passage from Europe to Asia by sailing north to the Arctic.

A British firm, Muscovy Company, sought out Hudson in 1607 and hired him to search for a northeast passage to China. It was thought at the time that if a ship sailed north just before the time when the North Pole receives three months of non-stop sunshine that this bright sun would melt the ice and the ship could then travel across the top of the world to the Spice Islands. The British, Dutch, and Spanish were all vying for routes, so these were important journeys.

Hudson was top pick in 1609 when the Dutch East India Company was hiring someone to find an eastern route to Asia to increase their share of the spice trade. They hired Hudson and told him to sail north toward the Arctic but to veer around toward Russia, where he should be able to find a route to the Pacific to take him to the Far East. Unfortunately, Hudson encountered so much ice that he was unable to get through.

Hudson was not going to be dissuaded from finding a passageway. As other travelers drifted back from North America, Hudson heard talk from them of the possibility of a southwest passage. He decided the thing to do was to explore for a southern route. Sailing across the Atlantic in his vessel, the *Half Moon*, Hudson made it to Chesapeake Bay and entered Delaware Bay on August 28, 1609. But he quickly realized that these waterways were not going to lead to the Pacific. He sailed north to New York Harbor and sailed up what is today the Hudson River. Trading with native tribes along the way, Hudson acquired beads, furs, and shells; he made it as far as Albany, New York, but the river narrowed at that point and he was forced to go back, once again realizing he had failed. (No route north of the Strait of Magellan existed to cross to the Pacific until the construction of the Panama Canal between 1903 and 1914.) However, Hudson's voyage established Dutch claims to the area, and New Amsterdam (Manhattan) became the capital of New Netherland in 1625.

On his return to Amsterdam, Hudson was detained by the British in England, and though he must have been angered by their demand to see his log books, by 1610 he was sailing under the English flag with funds from the British East India Company. This trip was to Iceland and Greenland, and then he sailed south to the Hudson Strait at the northern tip of

Labrador. Hudson and the men thought they had found the opening of the northeast passage, and they sailed on into what we now know as Hudson Bay. By November they had become trapped by the ice, and the men moved ashore to build a structure so they could make it through the winter.

Like other explorers we have encountered, Hudson was a flawed leader. He is said to have played favorites with men on the crew—providing those he liked with more food than the others, and then insisting that all crew baggage be searched in case one of them was hoarding food. It did not make him well liked.

As the ice thawed during the spring of 1611, Hudson prepared to continue exploring, but his crew was ready to go home. Hudson had garnered particular resentment from Henry Green, one of the crew members, and Green and a couple of others decided they had had it. They seized Hudson, his son, and seven others, and set them adrift in Hudson Bay in a small open boat. One account says Hudson and the men were given food and some supplies, but this account may have been written with the thought that the crew might need to document in a good light what happened to Hudson and the others, if the case ever went to court. As it happened, Hudson and the others were never heard from again, though in the winter

Henry Hudson and his men were abandoned by his crew and set adrift. (The painting erroneously states that Hudson was abandoned in 1610.)
Source: Prints & Photographs Division, Library of Congress, LC-USZ62-3923

of 1631–1632 another explorer found ruins of a shelter in the area that might have been built by them. Green and the other fellow who had led the mutiny didn't make it home either. They were killed in a fight with Eskimos.

Henry Hudson was brave, headstrong, and difficult. While he violated some of his sailing agreements and played favorites with the crew, tarnishing his ability as a leader, he extended the exploration of Canada and North America, and his discoveries formed the basis of the Dutch colonization of the Hudson River and later of English claims to much of Canada.

René-Robert Cavelier de La Salle (1643–1687): He Claimed Louisiana for France

René-Robert Cavelier de La Salle was a French explorer who set out for Canada in 1666; he eventually sailed down the Illinois and Mississippi rivers and claimed the region (the land along the Mississippi and its tributaries) for King Louis XIV of France. He named it "Louisiana" in the king's honor.

La Salle started his Canadian adventure as a fur trader. He often traded with Native Americans who told him of the great rivers—the Ohio and the Mississippi—that flowed into the ocean. La Salle began to think this might be the passage to the west (to lead to the East) that everyone was still looking for. For several years he explored on his own, and then in 1674 he returned to France for permission and funding. The king provided funding for La Salle to establish a Canadian trading post in what is now Kingston, Ontario. After an additional trip to visit the king, La Salle came back to explore and claim the Great Lakes, and then ventured down the Illinois River, continuing on down the Mississippi. In 1682 he buried an engraved plate at the mouth of the Mississippi and declared that the land belonged to France.

After yet another trip to France in 1684, La Salle set out with four ships and three hundred colonists to return to the mouth of the Mississippi. But the trip was ill fated: The weather was bad, their navigation was off, they were attacked by pirates, and hostile Indians greeted them onshore. One of the ships was taken by pirates in the West Indies, a second sank, and a third ran aground. They finally came to shore near what is now Victoria, Texas, and La Salle took a group on foot to try to locate the Mississippi. In 1687, some of the men mutinied, and La Salle was slain by the men.

This voyage was brought to life in 1995, when La Salle's ship, the *La Belle,* was located in twelve feet of water in a bay off the Gulf of Mexico. The *La Belle* had been transporting people who were to set up a colony at the mouth of the Mississippi. The search for the ship actually began in 1977, when the Texas Historical Commission asked a researcher in Paris to investigate any French ships that had sunk near Texas. Eventually

René-Robert Cavelier de La Salle onshore while supplies are unloaded from his ship. Source: Jan Van Vienen, Prints & Photographs Division, Library of Congress, LC-USZ62-3283

some historic maps were found to help assess where the *Belle* might have gone down, and the focus became Matagorda Bay, Texas. In 1978 a ten-week search took place, and promising evidence was found; but then for seventeen years no more funding was available, so no progress was made. Then in 1995 a new expedition was funded and a diving team recovered a distinctive bronze cannon bearing the crest of Louis XIV. They soon realized it was the *Belle*. During the summer of 1996 and through the spring of 1997 the vessel was excavated in its entirety and is currently undergoing conservation. The *Belle* is the oldest French colonial shipwreck found in the New World and is considered one of the most important archaeological finds of the twentieth century. Over a million artifacts have been recovered from the shipwreck.

After the excavation of the ship, the French government filed a claim for the ship and its contents. International law gives ownership of recovered ships to the country whose flag the ship flies. While many scholars feel the ship was actually La Salle's, not the French king's, this could never be proved. After a long negotiation, the official title to the wreck and its artifacts was given to the Musée National de la Marine in Paris, but day-to-day control was granted to the Texas Historical Commission in perpetuity. Many of the artifacts will eventually be on display at the Bob Bullock Texas State History Museum; others are on view at the local museum in Palacios, Texas.

Captain James Cook (1728–1779): Admired by All

Captain James Cook (who is almost always referred to as Captain Cook) was a British explorer who was a skilled cartographer and an excellent navigator. He began his career serving in the British navy, and during the siege of Québec (1759), he mapped out Newfoundland and explored much of the entrance to the St. Lawrence River. He gained great renown as a naval officer, was hired to command three separate voyages to the Pacific Ocean, and was the first European to make contact with the Hawaiian Islands and the eastern shore of Australia and to circumnavigate New Zealand.

A remarkable aspect of Cook's legacy was the thought he gave to keeping his men healthy. Though scientists did not yet understand the body's need for the vitamins contained in fresh fruits and vegetables, Cook seemed to understand instinctively that these were important. He insisted his men eat onions and pickled cabbage every day, and though it was difficult, he attempted to maintain stocks of fresh fruit and vegetables on board. As a result of these measures, Cook's men stayed relatively healthy and were less likely to come down with scurvy. He also ordered his men to bathe every day as well as to clean their clothes when they could and to air out their bedding.

Cook's Background

Cook was the son of a Scottish laborer who had moved his family to the Yorkshire area of England. As a teenager, Cook joined the British merchant navy and worked on a ship that traded at ports along the English coastline. Cook was a great student of experience, and he learned all he could about the tools of navigation, including the background needed (algebra, geometry, trigonometry, and astronomy) to properly navigate. As Britain began to gear up for the Seven Years' War (1756–1763), a series of multicontinent battles that pitted Great Britain and Prussia against France and Austria, Cook joined the British navy (1755) and was assigned to a ship that was to go to North America. In preparation for battle—what would be the siege of Québec City—Cook was assigned the job of understanding the "lay" of this new land so that the battle commanders could plan accordingly. He mapped the entrance to the St. Lawrence River and was the first to survey and map the entire coast of Newfoundland. His success at what he undertook brought him to the attention of the admiralty and the Royal Society (the national academy of science in the British empire), and this was to provide a new direction for his future.

Once the battles had subsided and the empire's attention turned to other interests, there were complementary areas that needed further examination: In the spirit of empire building, it was in the national interest to keep exploring new lands and new trade routes, and they were also beginning to see that greater scientific knowledge would make exploration easier. The Royal Society (the scientific academy) was thrilled to find a fellow who was smart, had a working knowledge of the science of the day, and was an excellent navigator. He factored perfectly into their plans for a voyage to the South Pacific in 1768. Cook was to pilot the *Endeavour*, with ninety-four scientists and crewmen aboard. The purpose of the trip was to observe and record the transit of Venus across the sun in order to determine the distance from the earth to the sun. This was so important to scientists that in an early example of international collaboration, several countries were sending out teams to observe the transit of Venus from several locations. This would permit scientists to come up with a more accurate measure using parallax (by placing observers in different locations, they could combine their results to increase accuracy). The overall intent was to come up with a correct figure for what we now refer to as the "astronomical unit" (or AU), which is now recognized to be about 93 million miles or 150 million km and is the distance of the earth from the sun. ("AU" is used to describe measures throughout the solar system.)

This voyage had to be carefully timed, as the pattern of these transits repeats every 243 years, with pairs of transits 8 years apart and separated by long gaps of 121.5 years and then 105.5 years. Johannes Kepler (1571–1630), the German astronomer who discovered that the earth's orbit was elliptical, predicted a transit in 1631, but his calculation was not accurate enough for anyone to see it. A British fellow named Jeremiah Horrocks observed the transit in 1639 from his home in Much Hoole, England. Horrocks was able to estimate the size of Venus and came up with a good estimate of the distance of the sun from the earth based on the information he obtained. Scientists wanted to check these calculations from multiple locations when Venus made her next transits in 1761 and again in 1769.

As was typical for scientific voyages of the day, Cook was accompanied by other scientists, who were going to document whatever they came upon. Cook left England in the *Endeavour* in 1768. They sailed southwest, going around Cape Horn (the southernmost tip of South America), and traveled to Tahiti where they were to wait for the passing of Venus. Among those on board was Joseph Banks, a botanist from a wealthy family, who collected over three thousand plant species on this journey. One of Banks's colleagues wrote to Carol Linnaeus, the Swedish botanist who established the system for botanical nomenclature, that Banks had supplied Cook's ship with £10,000 worth of his own equipment. It was thus described:

> No people went to sea better fitted out for the purpose of Natural History, nor more elegantly. They have got a fine library of Natural History: they have all sorts of machines for catching and preserving insects; all kinds of nets, trawls, drags and hooks for coral fishing; they have even a curious contrivance of a telescope, by which, put into the water, you can see the bottom at a great depth, where it is clear. They have many cases of bottles with ground stoppers, of several sizes, to preserve animals in spirits. They have several sorts of salts to surround the seeds; and wax, both bees wax and that of myrica; besides there are many people whose sole business it is to attend them for this very purpose. They have two painters and draughtsmen, several volunteers who have a tolerable notion of Natural History. . . .

Joseph Banks never made another voyage as significant, but he made a name for himself by bringing back at least eighty new species of plants and cataloging and publishing thirty-five volumes that outlined his discoveries, with artistic renderings by Sydney Parkinson, who had accompanied Banks and Cook on the journey. This added another contribution to the overall accomplishments of Cook's journey.

The men on board the *Endeavour* stayed in Tahiti until Venus had passed, and then Cook was told to open a sealed envelope from the Royal Navy with a second set of instructions for him after he successfully completed his mission for the Royal Society. Even as late as the eighteenth century, there were still many questions about the continents—no one knew for sure what was in the southern hemisphere, and when Cook opened the envelope, the request was for him to more fully explore the southern hemisphere. Mapmakers from the prior century (ca. 1570) believed that there were two major continents at each of the earth's poles, and explorers were regularly sent out in search of these lands. Since the *Endeavour* was traveling to the southern hemisphere anyway, this was a perfect assignment for Cook, who was regarded as a superior navigator and cartographer. He sailed around and mapped New Zealand, then continued west to Australia (1770), becoming the first European to visit the eastern shore. The *Endeavour* ran aground and was badly damaged on the Great Barrier Reef along the northeast coast of Australia. Cook and his crew stayed in the area for seven weeks while the ship was repaired. Cook then continued west and sailed back to England via the Cape of Good Hope (southern Africa). His circumnavigation of New Zealand and his estimation of the size and location of Australia had removed these landmasses from consideration, so the world was no wiser about the whereabouts of the "massive" southern continent when Cook's trip concluded.

When Captain Cook arrived home, he made his journals available for publication, and this raised his standing among the public. In 1772 the Royal Society commissioned him to search again for the "great continent," so Cook returned aboard the HMS *Resolution*, accompanied by another ship, the HMS *Adventure.* Cook sailed south, rounding Africa at the Cape of Good Hope. This time he charted for the ships to remain at a high latitude, and as a result, they were one of the first to cross the Antarctic Circle (January 17, 1773). His route actually took them within a very short distance of the mainland of Antarctica. They spotted an iceberg and followed along an "almost endless pack of ice." But they had to resupply, so they eventually turned north to Tahiti without ever definitively seeing land. During this journey, Cook visited the Friendly Islands, Easter Island, Norfolk Island, and New Caledonia. When the boats returned to England in 1775, Cook assured the Royal Society that there was no major continent in the most southern part of the hemisphere.

As was the custom of the day, Cook brought back "oddities" from his travel. As other captains had, Cook had picked up a native to help with navigation. (Given the distances the European ships could cover, these natives were often ill prepared to help navigate, though they were sometimes helpful translators.) Cook brought back "his" native Tahitian, Omia, to show around London much as one would display a museum curiosity. Omia became quite a novelty around London.

One navigational improvement used by Cook on his second voyage was a chronometer. With it, Cook and crew could more accurately keep track of his longitudinal position, something that they could not do well previously. (See chapter 5 for more information on how they navigated.)

Cook came home a hero. He was honored by the admiralty and made a Fellow of the Royal Society. While he could have retired, Cook was drawn back for a third voyage (1776–1779), this time to locate a northwest passage that would shorten the travel time by cutting across North America to Asia. Cook's theory was that it might be better to search from the Pacific side of the continent. Cook was to travel to the Pacific sailing east, while another ship was to travel to the Pacific from the west.

Aboard the *Resolution* again, Cook planned to take Omia home, so an early stop for Cook was a return to Tahiti. Next Cook explored the west coast of North America, all the way to Alaska as far as the Bering Strait. On this single visit Cook charted the majority of North America's northwest coastline for the first time, permitting him to determined the size of Alaska and gauge the distance to Russia. He discovered the Bering Strait was impassable and turned south, ultimately stopping in what he referred to as the Sandwich Island (after the fourth earl of Sandwich, who had partially financed his trip)—what we now know as Hawaii.

After an initial friendly visit and a stay in the Hawaiian Islands, Cook's ship ran into a problem and he had to return to the Kealakekua Bay area for repairs. Upon his return, the Hawaiians were less warm about providing food and aid to the men again. After a few days of squabbles, Cook attempted to kidnap their ruler, and Cook and several of his men were killed in the melee that followed.

Despite the ill will for Cook during these days, he was still held in high esteem by many Hawaiians, and he was given an honorable burial. (This may have involved roasting and eating part of him, but part of him was given back to his crew for a formal burial at sea.) His men made another attempt to get through the Bering Strait but eventually sailed home.

Because of the dozen years Cook spent exploring the Pacific Ocean, he provided Europe with more accurate information than they had had previously. Today we also recognize Cook's brilliant deduction that the Polynesians had traveled east from Asia, which even now seems like a nearly impossible feat.

Cook was a well-acknowledged notable of his time. Benjamin Franklin issued specific instructions in 1779 regarding Cook. Even though the American colonies were at war with Britain, Franklin instructed American captains that if they encountered Cook, they were to leave him alone, not delay him, and not plunder his goods. (Only later did Franklin learn that Cook had already met an untimely end.)

Cook's demise brought to a temporary standstill the British explorations of the South Pacific. But whalers and those who invested in them read of Cook's accounts of the abundant ocean life, so the channels to the Pacific were next filled with whaling ships.

John Ledyard (1751–1789): Bravery Personified

John Ledyard is certainly not a household name, and it is rare to find him in common geography books. But he was a brave American, a contemporary of the founding fathers, and representative of the spirit of the day. As a result of his efforts, he heightened the world's knowledge of geography.

Ledyard took a job on a ship that traveled to the Barbary Coast and to the Caribbean, but in England he left the ship and enlisted in the British navy. This put him in position to be among the crew of Captain Cook's third and final voyage to the Pacific, when Cook was determined to find a northwest passage from Europe to Asia. Ledyard maintained journals while on the voyage, though he had to turn them in to the British after the trip.

When the ship returned to England, Ledyard was still in service to the British navy, and he was assigned to a ship that was going to the colonies to fight against the Americans who were revolting. Once in his native land, Ledyard deserted and began to write down all he remembered about his travels. In 1783, Ledyard's *Journal of Captain Cook's Last Voyage* recorded much of what happened on that journey, and his description was the only eyewitness report of the murder of Cook and his men. His writings are considered a very valuable contribution to the history of exploration.

His description of their reception in Hawaii is also enlightening: He wrote that the Hawaiians who approached Cook's ships in canoes "appeared inexpressibly surprised, though not intimidated. They shook their spears at us, rolled their eyes about and made wild uncouth gesticulations." They also expressed bafflement at Ledyard's pointed military hat and wanted him to remove it so they could see the actual shape of his head.

Ledyard's next ambition was to find backers for what he predicted could be a very lucrative fur trade on North America's Pacific coast. In Paris, he met with Revolutionary heroes John Paul Jones and the Marquis de Lafayette as well as Thomas Jefferson, who was then serving as ambassador to France. No one provided him with money, but Jefferson encouraged him to become the first person to travel North America from west to east. For some reason, Jefferson suggested—or perhaps Ledyard decided on his own—that he should do so by first walking across Russia, crossing to North America at the Bering Strait, and then continuing southeast all the way to Virginia. He set out from St. Petersburg in September 1787. Though a few men started the journey with him, his only planned companions were two hunting dogs. Six months later, he reached Irkutsk (in south-central Russia, two-thirds of the way across the country) where he was arrested, accused of being a spy, and banished from the country.

Ledyard was not discouraged. He traveled to London where he connected with the African Association, which was then recruiting explorers for Africa. Ledyard suggested an exploration from the Red Sea to the Atlantic, but when his travel was barely under way, Ledyard came down with a stomach ailment in Cairo. He self-medicated, taking sulfuric acid and tartar emetic, and died in January of 1789 at the age of thirty-seven, probably from the medication.

History professor Edward G. Gray, who has helped bring Ledyard out of obscurity by writing a book about him (*The*

Making of John Ledyard: Empire and Ambition in the Life of an Early American Traveler, published by Yale University Press), notes: "His was an extraordinary life. Not only his adventures, but also his contacts with many great men of his time, made him a unique figure in early American history. But he was also representative of a particular type of person, a person whose opportunistic attitude helped extend European domination around the globe."

David Livingstone (1813–1873): He Opened Africa

We've all heard the phrase, "Dr. Livingstone, I presume?" and we have a vague idea that Livingstone explored Africa, but what really happened that spurred the fame of those words?

David Livingstone was born in 1813 near Glasgow, Scotland, and he first went to Africa at the age of twenty-seven with the intent of spreading Christianity. He intended to travel as much as possible, telling people about Christ. Several years later (1845), he married a woman named Mary Moffat, and she and eventually their children traveled with Livingstone early on. Over time, Livingstone made numerous trips across Africa during three different time periods: 1852–1856, 1858–1864, and 1866–1873. He was the first European to see the awe-inspiring Victoria Falls and the first to cross the entire southern part of the continent. When he went back to England, he wrote about what he found and gave lectures and became very well known. In 1866 he returned to explore more of the Nile River, but after he left on this journey, no one heard from him—though it was later learned that he was simply pressing on in further exploration. In 1871 a New York reporter named Henry M. Stanley led an expedition in search of him, eventually finding him and supposedly uttering the famous words, "Dr. Livingstone, I presume?" Stanley found Livingstone was unwell, and he had with him medical supplies. Livingstone began to recover, and Stanley and Livingstone continued on together for a time. In 1872, Stanley was ready to go back to England, but Livingstone refused all pleas for him to leave Africa.

Henry Morton Stanley greeting Dr. David Livingstone. Source: The Illustrated London News, Prints & Photographs Division, Library of Congress, LC-DIG-ppmsca-18647

Two years later, Livingstone died while praying. The local people removed his heart and viscera and buried them in African soil, and then set about a nine-month land journey to carry Livingstone's body and personal items and papers to a vessel that could take them back to England. In April 1874 he was buried in Westminster Abbey. He will always be remembered as the first European who went to Africa and came back to tell about the customs, the geography, and the ongoing slave trade of the many areas he visited.

Amazing dedication and fortitude were primary characteristics of all these explorers. Without their pressing on to learn more about the unknown land ahead of them, the world would never have known what was "around the next bend or down the next pathway."

5

How They Navigated during the Age of Discovery

Imagine for a moment what it must have been like to leave home on a ship or by horseback on a journey that might last several years and rely on little more than the stars as navigation points. The question "Where are we and how quickly can we get home?" took on new urgency. Today with car and handheld navigation systems, we forget that several hundred years ago, people depended on such basic tools as almanacs and knowledge of the heavens (a science known as "cosmography") in order to determine where they were going. Courage and trust must have been important components for those who were brave enough to leave their native land.

Extensive travel was taking place by the fourteenth and fifteenth centuries, but the methods used were still primitive. Land travel was particularly difficult. While certain areas, such as China, were often visited via an overland route, this travel method was arduous and often unsafe. The traveling itself was very hard work—sometimes walking in areas where the load on the animal needed to be lightened—and roads, if there were roads, were likely to be rutted paths. In addition, land travelers were sometimes attacked by hostile groups who inhabited the areas through which they traveled. For that reason, the men who ventured out on sailing ships tended to be more successful; they became the early explorers who made a difference. The Phoenicians and Carthaginians were the very early sailors, and at the beginning of the Middle Ages, it was the Vikings.

Early Skilled Navigators

The Vikings dominated the seas from about 800 to 1100, navigating in sleek, speedy vessels known as "longships." The original design of these vessels was called the "drekar," which featured square sails and a dragon-headed prow, the classic look of the boat associated with these men. This particular design was ingeniously flexible. It was capable of sailing across an open ocean, but when they saw signs of an inhabited community, the men could switch to "rowing power" for hit-and-run attacks. The vessel was steered with a single side rudder and could maneuver through shallow surf and river estuaries, providing even more flexibility. Using drekar ships, the Vikings navigated well, extending their reach as far as northern England and eventually on to North Africa. Over time, the Vikings developed another type of vessel, the knarr, which could carry cargo across open oceans. This permitted longer trips, and there is indication that the Vikings took to conducting some trade with Iceland and Greenland; we now know that they made it as far as North America, a trip that would have been very difficult without supplies.

Knowledge of how the Vikings traveled has not been easy to come by. The Norse legends provided a rudimentary knowledge of these people, but then in 1880—almost eight hundred years after these people sailed the oceans—a Viking ship, almost completely intact, was discovered in the waterlogged clay of a royal burial mound that dates to about 890. Eighty years later archaeologists had another stroke of good luck. From 1957 to 1962, additional Viking ships were found in a fjord in Denmark. From the placement of the sunken vessels, archaeologists feel they were sunk by the Norsemen intentionally to prevent invaders from entering the fjord. Some years later, contractors working along Denmark's Roskilde's waterfront were excavating to expand a museum to house Viking ships, and they were stunned to find nine wrecked Viking ships. Tree-ring analysis of the wood dates them to approximately 1025. One of the ships was 119 feet long and had room for at least seventy-two oars, meaning it probably sailed with a crew of about one hundred.

Exactly how the Norsemen navigated is not fully understood. They did not have compasses, so they must have relied on the sun and the stars. One of the Norse legends contains directions from Norway to Greenland, and it seems they relied on the sightings of birds and whales as well as distant landmarks to signal their location in relation to land. One saga refers to a *solarsteinn* ("sunstone") that was used for navigation. Some scholars think this was feldspar, a mineral from Iceland that could be used to indicate the direction of the sun when clouds obscured the view. However, a great deal of cloud cover would have made its use impractical.

In today's world, a discovery by one group of people usually leads to copycat inventions by others who have similar needs. So why didn't the Vikings' skills at sailing affect other people of this time? People in other countries must have desired to expand trade and to learn more about how they could benefit from nearby lands. The answer lies in the fact that most countries were physically isolated and methods of communication were extremely limited. It has only been in the last 100 to 150 years that an accomplishment in one country might be communicated to another country within a few days' time and copied soon after. Now it's so likely and viewed as such a threat that countries carefully guard information considered proprietary, such as building a better craft for space travel. Once it is built and launched, people throughout the world are aware of what has been accomplished, and a competition generally ensues to equal or improve on what was built. But at the time of the Vikings this kind of instant communication and copying simply didn't exist.

Two examples of Viking ships (far left and far right), with a Roman vessel in the center. Source: *The Iconographic Encyclopedia of Science, Literature, and Art* (1851)

Other Shipping Advances

During the Middle Ages, sea travel was actually safer than traveling by land, but unfortunately, maps of the time lacked accuracy. If a ship got off course or was diverted from a familiar path, it was very likely that the vessel was lost. And that was just *one* of the problems. While the Vikings were marauding around the northern seas in longships, the medieval people in Europe were tooling around in ships that needed to stay in the relative calm of the Mediterranean, Baltic, or North seas. There were two primary designs—barcas and barinels—but both featured similar problems: In order to adequately fill the sails, a ship had to travel into the wind at a minimum angle of 67 degrees. Then once the ships caught the wind, it was very difficult if not impossible to turn them around.

In the fifteenth century, Portuguese shipbuilders brought about a change in the ability for men to explore by developing a new type of sailing vessel. They developed the carrack, a three- or four-masted sailing ship with a high rounded stern and a flat back, generally with an upper deck (aftcastle) from which to fire upon other ships. The carrack offered space for crew, provisions, and cargo, which meant that the vessel could self-supply for longer than previous ships could; the ship was high enough that it could not be attacked by a smaller boat; the multiple sails offered greater propulsion and therefore greater speed; and the multiple sails permitted maneuvering as needed. As better materials became available and were used for lighter sails and stronger ropes, the number of men required to sail the ship decreased, which was also a big step forward. During the sixteenth century, the carrack design was altered to create the galleon, a stable but often heavier version of the earlier design, which was used both as a warship and to transport cargo.

The caravel was also developed by the Portuguese. It was a lighter, smaller ship that was preferable for exploring coastlines and inlets where the carrack-style boats had more difficulty navigating. Though no remains of this type of ship have been found, it is thought to have been based on an earlier fishing vessel, which may in turn have been based on an Arabic or even a Greek sailing vessel. (This information is primarily based on the fact that the etymology of the word *caravel* can be traced back to these sources.) When equipped with big square sails, the caravel was very fast, and it became known as the best sailing vessel of its time.

Portugal's Prince Henry the Navigator (1394–1460) chose the caravel as the ship he wanted for journeying along the West African coast in the 1440s. Its speed and ability to sail windward

Crew Management Presented Great Challenges

While scholars have come to realize that educated men during this time understood that the world was round and that there was no danger of sailing off the edge into oblivion, this same level of understanding did not apply to the men who served as the ships' crews. These men were usually uneducated, lower-class fellows who needed work. Their dream was not to explore new lands—they just needed a livelihood that would let them get by. As a result of little education and the tall tales that are a part of every culture, ship captains often met with great challenges in simply managing the men. As was discussed in chapter 3, John Cabot had to turn back because the men mutinied at the very beginning of one of his voyages. Many of the men who sailed did believe the world was flat, so ship captains had to dispel this belief, or at least assure the men that they would not go far enough to sail off the edge. Another common fear was that the ship would encounter a section of the ocean that was boiling hot and could consume the ship. The ocean south of the equator was particularly feared for this reason.

As ships became equipped with magnetic compasses, this was a definite improvement in the ability to navigate (when captains understood how to use them—not always a sure thing). Even then, some captains kept the compass in a secret location because superstitious crew members suspected that the compass was being pulled by sinister forces, creating a panic among the men.

Sea monsters were also a big concern, and while today we know that these things are not true, imagine embarking on a voyage that might last a couple of years and sailing in the open ocean without any sight of land for two or three months at a time. No wonder the men were afraid. Any captain had major challenges in coping with men who sometimes backed away from the work on board when they became consumed with fears.

were helpful for explorers, but there was a major disadvantage: Its large lateen sail (a triangular sail hung on a long yardarm) required a good-sized crew to maneuver it. This presented a challenge, because on long voyages ships had difficulty carrying enough food and water for a large crew. During the fifteenth century, adjustments were made by ship designers in order that this ship could be better used for exploration, and indeed, Columbus's 1492 expedition is thought to have included at least two caravels. (The *Santa Maria* is thought to have been a carrack.)

The Life of the Crew and Their Shipboard Tools

Aboard the ship, a typical crew pumped bilge, cleaned the deck, worked the sails, and checked the ropes and cargo. They worked four-hour shifts, measured by eight turns of the half-hour *ampolletas* (sandglasses). Sand moved through the glass more quickly when it was heated, and it was not unheard of for a sailor to warm the glass to make the sand—and his shift—run faster. Off duty, the men had no assigned barracks and had to find a place to sleep. (Only the captain had private quarters.) Religion was important to them and they began the day with prayers and hymns and ended with religious services in the evening. They received one hot meal a day cooked over an open fire on the deck. They ate ship's biscuit, pickled or salted meat, dried peas, cheese, wine, and fresh-caught fish. They often died from disease, hunger, and thirst.

Before the invention of additional devices (described below), sailors would measure the altitude of Polaris as they left home port. To return, the captain would navigate north or south to bring Polaris to the altitude of home port, and then turn left or right to retrace their steps depending on which way they had gone as they left. Arabs used one or two fingers' width or an arm or an arrow held at arm's length to sight the horizon below and Polaris above, and judged the distance in this way.

The chip log and the traverse board were ingenious in their simplicity and could be used by illiterate sailors to keep track of the ship's progress while each man was on watch. The "chip log" was a knotted line with a wooden weight attached at the end that was used to measure speed. A sailor counted how many knots were let off the reel in the time allotted. Multiplying the average rate of a ship's speed by a fixed amount of time gave a rough estimate of the distance traveled. This is why it became customary to describe a ship's speed by "knots." (Columbus's ships covered approximately 150 miles per day.) Prior to this method, the speed of a ship had to be calculated by having a crew member throw a chip of wood over the side of the vessel. He was then expected to judge the speed of the ship based on how long it took for the ship to move away from the wood. To avoid exhausting the supply of wood on board, a length of twine was tied to the log, and the log was retrieved and used repeatedly. Marks were added to the twine to allow for a more accurate reading of the ship's speed.

The traverse board—a board with pegs attached—was used to record the speeds and directions of a ship during a watch. Being able to read was not necessary to keep an accurate measurement. The board, kind of like a game board of yesteryear, had two parts. The top section was created to mimic a compass and had eight concentric rings with one peg hole at each point of the compass, as well as thirty-two compass points just as were shown on the ship's compass. It was used to record the direction the ship was sailing. Eight pegs were attached to the center of the compass with strings.

The bottom part of the board kept track of speed using four rows of holes. Each column represented a certain speed, measured in knots. During each man's watch, the crew member was to note directional headings each half hour by inserting a peg in the top portion of the board that corresponded with the heading sailed during the last half hour (based on the reading from the compass). The insertion of the pegs going outward was sequential by half hour. After all eight rings had been used, the crew member's four-hour watch was complete and a new person would take over, starting again at the center of the circle.

On the bottom portion of the board, a crew member was to insert a peg to represent the speed sailed during the hour. The speed was measured using a knot log. If the speed for the

hour was ten knots, the sailor would place a peg in the first row at ten knots. During the second hour the speed for that time period would be noted in the next row.

At the end of the four-hour shift, the crew member would collect the information and use it to record the vessel's track. Then a new person would come on watch and start over again with the pegs.

Though the devices were primitive by today's standards, many of them were very practical. Understanding what their tools were heightens appreciation for all that these explorers accomplished.

Celestial Navigation

Early navigation generally relied on sighting well-known landmarks. This meant that safety-conscious mariners needed to remain in close proximity to land. Stars were also helpful in navigating. Celestial navigation uses angular measurements (sights) between the horizon and a commonly known celestial object such as the sun, the moon, certain stars, or the planets. This permitted navigation when landmarks were not in view—but clear skies were necessary.

Celestial navigation can provide direction for which way to sail a ship, if it is time to sail home. But as they studied, explorers learned that celestial navigation could also be used to say "where we are now." At any given time, a celestial object (the moon, Jupiter, a particular star) is located directly over a particular geographical position on earth. This position is known as the object's "ground position," and almanacs were devised to record the "celestial coordinates" (latitude and longitude) for various objects at a particular time. The navigator could then use a sextant (see below) to measure the altitude of the celestial object, and use Greenwich Civil Time and the measured altitude to determine his position. Before the sextant was invented, navigators would hold their hand above the horizon with their arm stretched out. (The width of a man's finger is an angle just over 1.5 degrees.) One of the problems they encountered with celestial navigation was that as they traveled closer to the equator, the North Star dropped closer to the north horizon, making it more difficult to fix exact latitudes. Under these circumstances, they made use of the sun. Latitude could be measured as "noon sight" or from Polaris (the North Star). (Polaris always stays within 1 degree of the celestial North Pole.) If a navigator measures the angle to Polaris and finds it to be 10 degrees from the horizon, then he is on a circle at about 10 degrees north of the geographic latitude. Angles are measured from the horizon because locating the point directly overhead is difficult. Latitude could also be determined by the movement of the stars. If the stars rise out of the east and travel straight up, then the person observing this is at the equator. If they drift south, the observer is north of the equator, and vice versa. Longitude used to be measured primarily by the transit of the moon, since the earth turns about 15 degrees per hour, making it difficult to compare measurements day to day. Good measurements of longitude did not occur until the 1700s, when the chronometer (see sidebar) came into use.

For accurate maps, navigators needed latitude and longitude. Over time they needed more accurate instruments, and the astrolabe and sextant were among the items created (see below). For centuries, navigators could compute latitude by

Cook and the Chronometer

A chronometer is a timepiece with a special mechanism for ensuring and adjusting accuracy. It is used in determining longitude at sea or for any purpose where the very exact measurement of time is required. During Captain Cook's first trip, he gathered accurate longitude measurements, working with an early chronometer devised with another astronomer, Charles Green. They referred to the newly published *Nautical Almanac* tables via the lunar distance method and measured the angular distance from the boat to either the sun during daytime or one of eight bright stars during nighttime to determine the time at the Royal Observatory in Greenwich. They then compared that to the local time to determine direction via the altitude of the sun, moon, or stars. By his second trip, Cook had a K1 chronometer made by Larcum Kendall. It was the shape of a large pocket watch, and it copied the features of the clock that John Harrison had created for the ship *Deptford.* Its first test was on a trip to Jamaica in 1761–1762, and it kept accurate time.

measuring the angle of the sun or a star above the horizon with an instrument such as a backstaff or quadrant. Longitude was more difficult because it required precise knowledge of the time difference between points on the surface of the earth. The earth turns a full 360 degrees relative to the sun each day. Thus longitude corresponds to time: 15 degrees every hour, or 1 degree every four minutes.

The Almanac

An almanac is an updated record and reference tool that contains pertinent information in table form, usually arranged according to calendar dates. Astronomical data and various statistics are also found in almanacs, such as the times of the rising and setting of the sun, the moon, eclipses, and hours of the tides. Before there were maps, cartographers provided these types of charts so that navigators could establish latitude and longitude.

The need for this type of record and reference tool dated to Babylonian times; it was further advanced by the Greeks, who developed ways to keep track of the positions of the stars and the changing of the seasons. The earliest known modern-type almanac was created in the Islamic world by Abū Ishāq Ibrāhīm al-Zarqālī (Latinized as Azarqueil) in 1088; his guide provided the true daily positions of the sun, moon, and planets for a four-year period from 1088 to 1092.

By the Age of Discovery, this was a key reference tool for any traveler, including those on shipboard. Updated regularly, an almanac provided astronomical or meteorological information, usually including future positions of celestial objects, star magnitudes, and culmination dates of constellations. In 1457, the first printed almanac was published at Mainz by Gutenberg.

The people of the late Middle Ages and the early part of the Renaissance also believed in foretelling the future, so the almanac was a helpful tool to those who were making predictions. Today we have these types of almanacs as well as a wide range of expanded almanacs containing everything from sports tables and statistics to political almanacs with complete records of elections past.

The Compass

The first compass is thought to have been invented by the Chinese, probably during the Qin dynasty (221–206 B.C.E.), when they discovered that lodestone (a mineral composed of an iron oxide) automatically aligns itself in a north-south direction. They began to use the compass to "align" with the forces of the earth to help them maintain balance in their lives. By the eleventh century, they were using it as a direction finder, but they referred to the device as a "south-pointing fish." For hundreds of years now, people have been fixated on locating "north" with a compass, but the original interest obviously was exactly the opposite. The use of the compass for navigation seems to have occurred in China within a few years of this time (the eleventh century). In a book written in 1119, *Pingzhou Table Talks*, a compass is referred to as being used for navigational purposes.

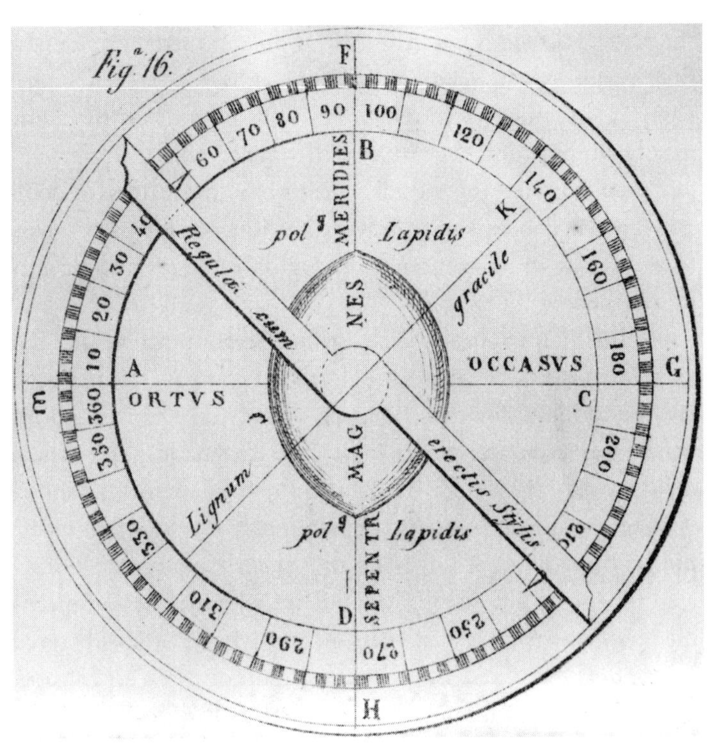

A mariner's compass from the 1200s. Source: Prints & Photographs Division, Library of Congress, LC-USZ62-110320

Magnetic compasses work because the earth itself acts as a bar magnet, possessing a north-south field that causes freely moving magnets to take on north-south orientation. The earth's magnetic field is not quite parallel to the exact north-south axis of the globe, but it is close enough to make a compass a reasonably good guide. (The rate of inaccuracy varies from point to point on the earth and is known as "variation" or "declination.")

Over time, people began to develop ways to create magnetic compasses. If lodestone itself is floated in a liquid, it will align itself to point to the polestar. Shortly thereafter, people must have realized that if an iron needle was touched by a lodestone, it could be magnetized so that it would align itself directionally. If a magnetized iron needle is mounted on wood in such a way that it "floats," then the needle will align itself with the earth's magnetic field, taking a more or less north-south direction. Once "north" is identified, of course, other directions can then be determined.

The English, who relied on being able to travel to maintain their empire, made many improvements to the magnetic compass. At first only north and south were marked on the compass bowl, but soon they developed ways to note thirty principal points under the needle, permitting navigators to look down at the compass to tell their direction. Keeping a compass level was also important, and the English came up with devices that helped with this, too.

In the fifteenth century, navigators began to realize that the compass did not point to true north, so compass needles were created to allow for this, pointing slightly east of true north, which was accurate for the day. (The magnetic poles of the earth actually shift; today's compasses now allow for this shift.)

The use of the compass did not attract the attention of explorers until about 1300, when the mariner's compass was created for use on shipboard. It is unclear whether the Europeans hit upon the discovery on their own or whether they observed its use in China and brought the invention back to Europe. (China used a needle floating in water for its direction finder, while Europeans created a "dry compass" mounted on a board but floating free so that it could pivot.) The compass revolutionized travel. Navigators who had relied on the stars could now establish their position beneath cloudy skies. European navigators eventually traveled around the world with the aid of this navigational device.

The use of the compass also improved the economics of sailing by extending the months when ships could be afloat. From ancient times, European ships had stopped sea travel from October to April, partly because the men couldn't depend on having clear enough skies to navigate during those months. It was expensive and dangerous to be waylaid on a journey. The use of the compass meant that by 1290, sailors began starting out in late January or early February, and this extended their ability to reach different areas and get back before the end of the year.

The Modern Compass

Modern compasses use a magnetized needle or dial inside a fluid-filled capsule (oil, kerosene, or alcohol is common). The fluid dampens the movement of the needle and causes the needle to stabilize quickly rather than oscillate back and forth around the magnetic north.

A gyrocompass is a compass that finds true north by using an electrically powered fast-spinning wheel and friction forces in order to exploit the rotation of the earth. Gyrocompasses are widely used on ships because they find true north as opposed to magnetic north (the direction of the earth's rotational axis); they are not affected by ferrous metal in the ship's hull. They are even used sometimes in preference to a Global Positioning System (GPS). They require no energy supply, and while a GPS can be affected by things that block the electronic signals, the gyrocompass can operate independently of that.

More accurate devices now exist for directional information. The astrocompass and the gyrocompass can both be used to find the earth's true north, as opposed to magnetic north. These forms also are not influenced by nearby electrical power circuits or large masses of ferrous metals.

A recent development is the electronic compass, which detects the magnetic directions without requiring moving parts. This device frequently appears as an optional subsystem built into GPS receivers.

The Mariner's Astrolabe

This device was used to understand the time and positions of the sun and the stars and was a very important tool for astronomical studies. Astrolabes reflected how the sky looked at a specific place at a given time. The sky was drawn on the face of the astrolabe and then positions in the sky were marked so that they were easy to find. There were movable components that could be adjusted to reflect a specific date and time. The one used by mariners was a specially designed astrolabe that was marked with rings and was used for measuring celestial altitudes.

The Cross Staff

By the sixteenth century, this was an important navigational instrument on shipboard. The implement consisted of a long staff with a perpendicular vane that slid back and forth along it. The staff was marked with graduated measurements computed by trigonometry. Angles of a distance could be measured by holding the staff up so that the ends of the vane were level with the points to be measured. It could be used to measure anything from the height of a building to the topographical features of a hill or the angles between stars.

As originally invented, the cross staff had one vane and was very long. Its length made it difficult to handle on a rocking ship. Mariners found that by adding more vanes of graduated sizes placed at graduated distances, they could reduce the length of the staff to about 2.5 feet, calibrating the measurements specifically for use on shipboard. The cross staff was usually used to measure the altitude of the polestar above the horizon. If it was cloudy, then it was impossible to take this measure, so they attempted to measure the altitude of the sun. This necessitated looking directly into the sun, which damaged their eyes, so they soon developed a backstaff. This device worked similarly to the cross staff, but with the backstaff, the pilot faced away from the sun, sighted the staff on the horizon, and then adjusted a cursor until the shadow fell upon the sight through which the horizon appeared, so that the resulting arc could be measured.

The Sextant

The sextant was a key navigational instrument—a "point and shoot" device—developed to identify the angle between the horizon and a celestial body, such as the sun, the moon, or a star. From this calculation, sailors could then determine their latitude and longitude. (An earlier version of the sextant was the quadrant—it was used by Columbus. Quadrants had to be held exactly vertical in the plane of the heavenly body to get a good measure of Polaris, and some of its settings were subject to being off from the wind.)

The sextant is a triangular-looking instrument with one end arced and marked off in degrees. One of the arms is movable and pivots from the center of the circle. A telescope, mounted to the framework, lines up with the horizon. The movable arm has a mirror mounted on it, which reflects the star into a half-silvered mirror in line with the telescope, and to the person looking through the telescope it brings it so that it appears to coincide with the horizon. The angular distance of the star above the horizon is then read from the graduated arc of the sextant. From this angle and the exact time of day as registered by a chronometer, the latitude can be determined within a few hundred meters (looking at published tables).

The name "sextant" comes from the 60-degree span of the arc, or one-sixth of a circle. The first devices were octants (from the 45-degree span of the arc, one-eighth of a circle), but sextants were developed to calculate longitude from lunar observations. They replaced octants by the late 1700s.

Though Sir Isaac Newton (1643–1727) developed the principle of a double-reflecting navigational instrument, his idea was never published. Working separately, British mathematician John Hadley (1682–1744) and Thomas Godfrey (1704–1749), an optician in Philadelphia, independently developed it.

The navigational sextant offered advantages of accuracy and both day and nighttime use. (The backstaff is difficult to use at night). When adjusted properly, the horizon and celestial object remain steady when viewed through a sextant, even when aboard a moving ship. (Because it views the horizon and

the celestial object through two opposing mirrors, motion does not affect the measures.) It is not dependent on electricity.

While there were many advances in sailing technology during these years, the machinery was still quite primitive by today's standards. If you think of departing from a dock on a sailing ship with only the aforementioned devices, it is all the more remarkable that these gentlemen sailed off and were able to return home, too.

PART TWO

The Nuts and Bolts of Geography

6

How the Earth Works

One of the major issues that scientists wrestle with concerns when the earth came to exist and why we have the topography we do. Both theories are constantly being reexamined and updated based on the latest scientific discoveries. What we do know is that the world has been constantly changing, and nowhere is this more poignantly demonstrated than at the site of the World Trade Center in lower Manhattan. In preparing the foundation for Tower 4 of the new World Trade Center, engineers must understand the rock contours in order to plan out solid concrete footings for the building. This has led the workers to uncover topography that includes steep cliffs, vast holes, and surprising variations in the steel-gray bedrock, all part of a landscape created by the glaciers that covered the area at least twenty thousand years ago.

This chapter discusses what is currently known about the "when" and the "how" of the earth and its topography.

How Old Is the Earth?

At this point, there is no way to pinpoint the age of the earth exactly because it is thought that the earth's oldest rocks have been recycled and destroyed by the process of plate tectonics (discussed later in the chapter). However, scientists have been able to determine the probable age of the solar system, and from there they have calculated an age for the earth by assuming the earth and the rest of the solid bodies in the solar system were formed at the same time. (It is actually easier to date the moon because more of its ancient rocks are available, since it does not have tectonic plates to destroy the older

rocks.) Though scientists have found rocks on the earth as old as 4.03 billion years old (northwestern Canada), it is thought that the earth is actually about 4.6 billion years old, which would be consistent with current calculations of 11–13 billion years old for the Milky Way and 10–15 billion years for the age of the universe.

But keep following the news for updates. In 1933 they thought the earth was only 2 billion years old. New scientific methods may change this "birthday" calculation yet again. (Any shifts or changes to the time scale are overseen by the International Commission on Stratigraphy.)

How Do We Know How Old Things Are, Anyway?

As scientists began to try to figure out how old things were, they first had to rely on coming up with a geologic time scale that provided some age ranges of when various forms of rock were formed. They then estimated the age of rocks and fossils and artifacts based on the estimated layer of where something was found. As technology improved, better methods have been created, and one that has been particularly useful is radiocarbon dating.

Radioactive carbon 14, an isotope that has become useful in dating ancient artifacts, was first discovered in 1940 by Drs. Martin Kamen and Samuel Ruben, both of whom were chemists at the University of California, Berkeley. The discovery of carbon 14 occurred at Berkeley when Kamen and Ruben were working to find a radioactive isotope of carbon that could be used as a tracer investigating chemical reactions in photosynthesis. (Almost all carbon atoms in nature are carbon 12, containing twelve protons and neutrons in the nucleus, so the identification of this radiocarbon isotope with six protons and eight neutrons was noteworthy.) In 1947 Willard Libby, a chemistry professor also connected with the University of California, Berkeley, was experimenting with different ways in which this isotope could be useful, and he discovered that carbon 14 could be used as a method for dating artifacts and human remains from long ago.

Over time Libby observed that during photosynthesis plants absorbed carbon 14, a radioactive isotope (one of two or more atoms having the same atomic number but different mass numbers). Throughout its lifetime, a plant absorbs a constant amount of carbon 14. When the plant dies, it stops absorbing this radiocarbon element. The rate of decay of the carbon 14 is predictable and measurable in objects as old as 45,000 to 50,000 years. This measurable rate of decay is known as the "half-life." Libby discovered that by determining the concentration of carbon 14 left in the remains of a plant, he could calculate how much time had passed since the plant

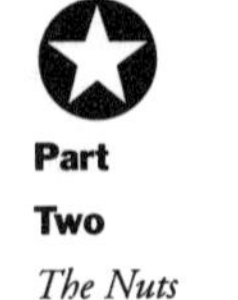

William Morris Davis, the Father of American Geography

While William Morris Davis is an almost totally forgotten figure in American history, he actually did contribute greatly to getting the field of geography off the ground in America. When you do read of him, he is described as "the father of American geography."

Born in Philadelphia, Davis graduated from Harvard at the age of nineteen, earned a master's degree in engineering, and spent three years in Argentina at a meteorological observatory. He returned to Harvard, where he became a professor of physical geography, teaching many of the great American geographers who were to follow him. (These fellows, too, are totally unknown today but contributed to our understanding of the world.)

One of Davis's contributions was the founding of the subfield of geomorphology, the study of the earth's landforms. At that time, the most common belief as to how the earth had formed was that the landforms that existed were what was left following the great biblical flood. Davis began to believe there were more scientific factors involved. He wrote on a theory that is commonly known as the cycle of erosion or the geomorphic cycle, which explained how mountains are created, mature, and then become old. Today this theory is viewed as simplistic, but in Davis's day, it was a very important stepping-stone to more knowledge.

Davis also founded the Association of American Geographers in 1904 and helped build it into an organization that had high standards when publishing work in the field. He also called for an increase in geography curriculum in universities and often wrote about how geography should be taught in order to make it interesting to as many people as possible. (He would have been very happy that books like this are published!)

died. Since animals and humans eat plants, the carbon 14 they contain also undergoes an even rate of decay upon their death. Because most artifacts are made from an organic substance, the date of the object can be judged based on the decay of the organic content. This finding has enabled more accurate estimates of the age of many artifacts and skeletons that formerly were dated through less reliable methods.

Because objects more than 45,000 to 50,000 years old do not have enough carbon 14 to measure, scientists have found they can use similar dating techniques by employing elements with a longer half-life (a longer time to decay) than carbon 14. Among the elements most widely used currently are potassium 40 with a half-life of 1.25 billion years, uranium 238 with a half-life of 4.5 billion years, and rubidium 87 with a half-life of 49 billion years.

Continental Drift and the Continents

As early as 1596 a Dutch mapmaker by the name of Abraham Ortelius developed the theory that at one time the continents of the earth fit together like a big jigsaw puzzle. He theorized that the continents might have been separated by earthquakes and floods. But science was too young at the time for Ortelius's theory to gain any backing.

In 1912 a young scientist named Alfred Wegener (1880–1930) began to develop a theory that became known as "continental drift." Wegener was a lecturer in astronomy at the University of Marburg, and he came upon a scientific paper that noted the existence of identical plant and animal fossils on two sides of the Atlantic. (In 1859, a French scientist, Antonio Snider-Pellegrini, had introduced the idea that all the continents were once connected. Other scientists began to locate similar plant fossils on various continents, which led them to consider the possibility that one huge landmass might have existed at one point.) Wegener began to contemplate this discovery as well as to contemplate the "puzzle piece" aspect of the continents. Others had also noted that South America looked as if it could nestle under part of Africa, and Greenland appeared to be a "connector" between Europe and North America.

Wegener determined that until the Carboniferous ("coal-forming") Period about 300 million years ago, the continents were a single supercontinent—he referred to it as "Pangaea" (from the Greek for "all the earth"). Wegener explained that Pangaea then split for some reason, and its pieces have been moving away from each other ever since. Wegener based his theory on several pieces of evidence, including geological, paleontological, and climatological factors:

- *The composition of the Mid-Atlantic Ridge, which forms such islands as Iceland and the Azores.* This seemed to Wegener to be material left behind when the continents that now flank the Atlantic broke apart. He also noticed how mountain ranges and glacial deposits match up when continents are envisioned together, forming a continental jigsaw puzzle.
- *The surprising distribution of fossil remains of trees and other plants that were around during the Carboniferous Period, which had been noticed by scientists before him.* Though each region of the earth seems to have its own specialized vegetation, botanists have found that some plants such as *Glossopteris* (the seed fern) thrive in widespread locations including India, Australia, South America, and South Africa, and also in coal seams in mountains near the South Pole.
- *Animal distribution.* Wegener felt that a land called "Lemuria" once linked India, Madagascar, and Africa, and this would explain the widespread distribution of the lemur and the hippo. That marsupials such as the kangaroo and the opossum are found only in Australia and the Americas made Wegener also link Australia with distant South America.

Drawing upon his knowledge of various aspects of science as well as his own explorations, he gave a talk on "continental displacement" to the Frankfurt Geological Association in 1912. By 1915, he was ready to write what would be the first of four versions of *The Origin of Continents and Oceans*. While Wegener continued to come up with evidence that backed his theory, he was unable to explain what would have pushed the continents apart. He finally concluded that the continents were like great barges, plowing their way to their current positions like icebreakers. Most scientists, however, found this explanation preposterous and dismissed the whole theory as a result.

In the autumn of 1930, Wegener, who was also trained as a meteorologist, agreed to accompany a scientist friend to help establish a weather station in Greenland. Though Wegener reached the intended destination despite horrific weather conditions, he died—it is thought he had a heart attack—when he left the station to go for more supplies. Otherwise, he might have lived long enough to see that scientists eventually embraced his theory.

Just prior to Wegener's death, British geologist Arthur Holmes became an active supporter of Wegener's theory and proposed that over a long period of time the heating and cooling of the earth (thermal convection) might be enough to break apart landmasses and cause the continents to move. But nothing about this idea was catching on. Holmes, too, was totally ignored until the 1960s.

Studies of the Ocean Floor Alter Theory

In the 1940s and early 1950s, geophysicist and oceanographer Maurice Ewing began taking seismic readings of the ocean floor. Ewing's readings and measurements were sent back to his research assistant Marie Tharp (1920–2006), who had been permitted to train as a geologist only because of the shortage of male workers caused by World War II. Working along with colleague Bruce Heezen, Tharp performed detailed mappings of the ocean floor. Their work eventually uncovered a 40,000-mile underwater ridge that encircles the globe. This geophysical data laid the foundation for the conclusion that the seafloor spreads from central ridges and that the continents are in motion, which paved the way for acceptance of the theories of plate tectonics and continental drift.

Today Marie Tharp is considered a pioneer of modern ocean-floor cartography, but it has only been since the mid-1990s that she has been recognized for her work. (She was not in anyone's science book, even though she deserved to be.)

In the 1950s scientists also started experimenting with magnetometers (adapted from airborne devices developed during World War II to detect submarines), and they began to notice something unexpected: The magnetic fields on the ocean floor changed direction periodically. Initially, scientists felt this was because basalt, an iron-rich volcanic rock that makes up the ocean floor, contains magnetite, which can distort compass readings. However, based upon the study of lava formations in Hawaii, scientists began to realize that as newly formed rock cools, the magnetic materials record the earth's magnetic field at the time. They saw that the earth's magnetic field reverses at intervals that range from tens of thousands to many millions of years, with an average interval of approximately 250,000 years. The last such event occurred some 780,000 years ago.

This new information gave scientists two new clues about the earth's history:

1. When molten rock rises from below the earth's crust (in this case, underwater) and hardens into new crust, it hardens with the magnetic pull of its own "present-day" polarity. As it pushes up, it pushes along the previous crust in a "conveyor-belt-like motion," so "young rock" is near the crest of the ridge, and the farther away you move from that ridge, the progressively older will be the rocks.
2. This information, combined with the magnetic "striping" on the seafloor, provided scientists with one more way to understand how the earth was formed. Stripes of rock parallel to the ridge crest alternate in magnetic polarity (normal-reversed-normal, etc.), suggesting that the earth's magnetic field has flip-flopped many times.

Working with the information that had been acquired by 1962, American geologist Harry Hess proposed the theory of seafloor spreading. If this was so, and the earth's crust was expanding along the oceanic ridges, then Hess concluded that it must be shrinking elsewhere—eventually (millions of years later) descending into oceanic trenches. Hess suggested that the continents did not float about, but interacted with the

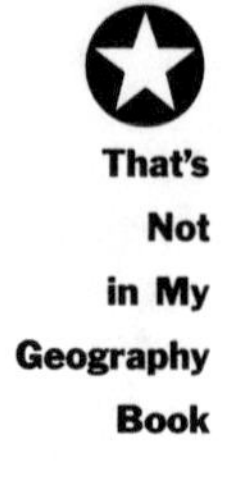

oceanic crust. Plate interactions formed mountain ranges, earthquakes, and volcanoes. Hess also proposed a mechanism that was driving the movement of plates.

By explaining both the zebralike magnetic striping and the construction of the mid-ocean ridge system, the seafloor spreading hypothesis quickly gained converts and represented another major advance in the development of the plate-tectonics theory.

How It Works

Today we know that the earth's surface is made up of eight to twelve big plates and twenty or so smaller ones. They all move in different directions at different speeds; some are slow moving (a fraction of an inch per year), while others are relatively speedy (a few inches per year) in their progress. You cannot necessarily figure out the plates by knowing the continents. For example, the North American plate roughly traces the outline of the western coast (where everyone knows there is a great deal of earthquake activity), but the eastern side of the plate extends halfway across the Atlantic to the mid-ocean region.

The main features of plate tectonics are

- The earth's surface is covered by a series of crustal plates.
- The ocean floors are continually moving, spreading from the center, sinking at the edges, and being regenerated.
- Convection currents beneath the plates move the crustal plates in different directions.
- The source of heat driving the convection currents is radioactivity deep in the earth's mantle.

Most of the world's active volcanoes are located along or near the boundaries between shifting plates and are called plate-boundary volcanoes. The peripheral areas of the Pacific Ocean basin, containing the boundaries of several plates, are dotted with many active volcanoes, which form the so-called Ring of Fire. The ring provides excellent examples of plate-boundary volcanoes, including Mount St. Helens. However, some active volcanoes are not associated with plate boundaries, and many of these so-called intraplate volcanoes form roughly linear chains in the interior of some oceanic plates. The Hawaiian Islands actually provide the best example of intraplate volcanoes, caused by the fact that the Pacific plate slides over "hot spots" that can cause volcanic activity.

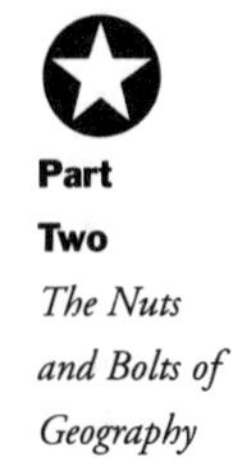

It's the Tectonics That Cause Earthquakes

Over time, scientists began to realize that the earth consists of two major layers. The outer shell, the lithosphere, is rigid crust, broken into fragments or plates. It floats on a soft, flowing inner shell called the asthenosphere. The movements of these plates reshape continents, build mountains and valleys, and affect the dominance and evolution of species. Because the sides of a plate are either being created or destroyed, its size and shape are continually changing. There are three types of boundaries that define the way the plates bump against each other.

Convergent boundaries. These consist of two plates colliding or pushing against one another. The fate of these collisions is a bit like the game "rock, scissors, paper," in that there is a predictable pattern of dominance. If a very dense oceanic plate encounters a less dense continental one, the oceanic plate is generally forced underneath, forming a subduction zone. These encounters create changes in the geology of the area that can result in the creation of mountain ranges and volcanoes. The mountainous spine of South America and the Cascade Mountains of North America are good examples of this.

Divergent boundaries. A divergent boundary consists of two plates moving away from one another. Over time, the space is filled with new crustal material from molten magma. An example of this is the oceanic ridge systems that cause ocean spreading, such as with the Mid-Atlantic Ridge.

Transform boundaries. These occur where two plates slide past one another. These are known as "slip faults." Plate-to-plate friction means they don't simply glide past each other. Stress builds up and when it reaches a level that exceeds the slipping point, there is motion along the fault, causing an earthquake. For example, scientists know that the Pacific plate

How Earthquakes Are Measured

Seismographs located all over the world measure the shaking of the earth. The measure of earthquakes we all hear about is the "Richter scale." In 1935 Dr. Charles Richter, a geologist at the California Institute of Technology, proposed using this method, which is a mathematical scale measuring the magnitude of ground movement. Like the waves created by a pebble tossed into water, the ripples of an earthquake weaken as they get farther and farther from the epicenter of the quake. Because the earthquake will affect different areas slightly differently and it's impossible to measure each spot where an earthquake is felt, the measurement of any earthquake is usually gathered from at least two different seismographs and a range of the magnitude is presented, such as from a low of 7.6 to a high of 8.5.

is moving north while the North American plate is moving south; this causes friction along the San Andreas fault along the Pacific Coast.

Earthquake Vulnerability in the United States

When we think of "earthquake vulnerability" here in the United States, we think of California. We've all heard about the San Andreas fault, and scientists feel that the fault is ready for a big adjustment some time soon. But while the focus may be on California, this is only part of our U.S. story. There are forty-one states and territories in the United States at moderate to high risk for earthquakes, and no region of the country is immune:

- Alaska experiences the greatest number of large earthquakes—most located in uninhabited areas. One of the largest was near Anchorage in 1964; it measured 9.2 on the Richter scale. In some places the ground rose thirty feet, and the earthquake set off a tsunami that killed 122.
- The largest earthquakes felt in the United States were along the New Madrid fault in Missouri, where a three-month-long series of quakes from 1811 to 1812 included

> three quakes larger than a magnitude of 8 on the Richter scale. These earthquakes were felt over the entire eastern United States, with Missouri, Tennessee, Kentucky, Indiana, Illinois, Ohio, Alabama, Arkansas, and Louisiana experiencing the strongest ground shaking.

That said, part of our thinking about California is correct; the state does experience the most frequent, damaging earthquakes. The state's San Andreas fault is almost an identical twin to the North Anatolian fault, which produced the magnitude 7.4 earthquake near Ismit, Turkey, in 1999, killing more than fifteen thousand with uncounted numbers buried in the rubble. In addition, scientists have for the first time confirmed that downtown Los Angeles is situated on what is known as a blind thrust fault, a type of fault capable of producing a devastating earthquake. (In a quake caused by a thrust fault, blocks of earth move diagonally, almost vertically. In a strike-slip fault like the San Andreas, the opposing earth plates slide past each other horizontally.)

As we learn through any news coverage of major earthquakes, a strong one can collapse buildings and bridges, disrupt gas, electric, and phone service, and sometimes trigger landslides, avalanches, flash floods, fires, and huge destructive ocean waves (tsunamis).

Often, the major devastation from an earthquake occurs from the events it spawns: The 1906 earthquake in San Francisco was serious, but it was the devastation of three days of fires that caused the most serious damage. Because of broken water mains, it was impossible to fight the fires. Twenty-eight

The 1906 earthquake and following fires devastated the city of San Francisco, California. Source: Pillsbury Picture Co., Prints & Photographs Division, Library of Congress, LC-USZ62-64303

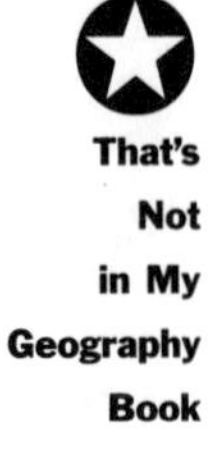

thousand buildings were destroyed; 300,000 people were left homeless, and the death toll was approximately 700.

While the shaking of the earth is frightening, it is seldom the direct cause of death or injury. Building collapses, flying glass, and falling objects are generally the cause of earthquake-related injuries. This is extremely instructive in making adequate preparations. You will likely be able to withstand the earth tremor; what you need to focus on is creating an environment that diminishes hazards from other risk factors such as falling objects.

Like all natural disasters, the seriousness of an earthquake is as dependent on the population of the region as it is on the magnitude of the actual disaster—an earthquake that is 8 or 9 on the Richter scale and occurs in a deserted area would not be viewed as seriously as one that measures a 5 or 6 but occurs in a major city. Time of day also makes a big difference. If most people are at home and asleep, fewer people will be hurt than if it's midday and people are in office buildings, on freeways, and otherwise moving about outdoors.

What we do know is that earthquakes are going to continue to shake our world, and the only question is where and when.

7

Mapmaking Then and Now

On September 11, 2001, the day of the horrific terrorist attack against the United States, one of the most needed items was a very old map. It was the map of New York City's water and sewer grid, which is kept in Queens in the Bureau of Water and Sewer Operations. As water mains burst during the horrific damage that hit lower Manhattan and threatened to flood the area and compound the already disastrous situation, emergency workers needed quick access to multiple shutoff valves in order to turn off the water as quickly as possible. One man, the deputy commissioner of the department of environmental protection, sat in Queens by a two-way radio pinpointing the locations needed by the on-site workers and telling them where to go.

This is just one type of vital story told by a map. Maps communicate spatial information acquired from measurement of space, present it in such a way that it can be depicted on a flat surface, and make it accessible. Maps used to have to be hand-drawn using symbols and lines to represent geographic phenomena. Today computers have revolutionized the mapmaking industry and increased our access to all sorts of commercial-quality maps, making dynamic interactive maps available in the home via computer.

Change through digitization is almost complete for the New York City sewer and water maps, too. After years of city workers and contractors trekking to Queens to look at browning paper maps, some of which date as far back as the Civil War (only dire emergency gave them access to that information via phone or radio), the city has been investing in a project over the last ten years that will not only digitize the maps but also connect them with other technical information such as the size of the pipes, the date they were last repaired, and what they are made of, all of which will be retrievable by the click of a computer mouse. (GPS technology ensures that the maps are accurate to within eighteen inches.) The workers gain a virtual snapshot of what is in the ground. This is a step forward in increasing productivity because it offers a way to retrieve information more quickly.

Maps show us the world as others have seen it, or as they want us to know it. They can be stunningly beautiful and fascinating to study. It is hard to resist a good map—whether it's an early view of the United States, where the Louisiana Purchase is shown as an empty "unknown," or a Google map of your own neighborhood. Maps are endlessly interesting.

Mapmaking Then

Maps have been key to exploring everything from the valley over the next hill to the underground world of modern subways. While much early exploration was done without accurate maps, one of the goals of almost any expedition was always to set down the facts about the geography of an area so that others could follow. If they had no opportunity to write information down, early explorers would mark the territory, with cuts on trees or by creating stone mounds so that people would know where to go.

The science of mapmaking—cartography—extends back to clay tablets on which ancient people noted the whereabouts of parts of the Mesopotamia area. The oldest surviving attempt to depict the world was painted perhaps 7,000 or 8,000 years ago, on a cave wall near Jaora, India. But the first great influence on the world of mapmaking was Claudius Ptolemy, a Greek mathematician, astronomer, geographer, and astrologer. Ptolemy developed the latitude and longitude coordinate tables that permitted mapmakers to project the spherical earth onto a flat piece of paper. Though none of his maps still exist, he wrote a work on mapmaking in 151 C.E. and called it *Geographia*. Ptolemy's ideas remained the basis for explorers and mapmakers alike until the tenth century, so understanding his impressions of the world explains much of the thinking of the very early explorers.

Ptolemy believed that the world stretched from Iceland and the Canary Islands in the west to Ceylon in the east, with a mass of unknown lands south of North Africa and beyond India. He thought Africa was connected to an undiscovered southern landmass, which would have kept ships from reaching Asia by sailing around Africa. He also put forward the belief that was followed well into the Age of Discovery—that the Eurasian continent was much wider, that the landmass of India was much closer, and that to the west of Europe lay India, with no continent in between. It took more than fifteen hundred years before the explorers realized that the lands they were exploring when they went west were what became known as the continents of the Americas, not India.

The Arabs improved on Ptolemy somewhat in the tenth century, when the scholar Massoudy suggested that a channel existed between southern Africa and an unknown landmass in the southern extremities of the world—meaning that sailors might be able to navigate around Africa.

During medieval times, mapmaking entered into a "dark" period. Geographers blended fantasy with factual maps, which led to less accuracy and a loss of information. Maps during this

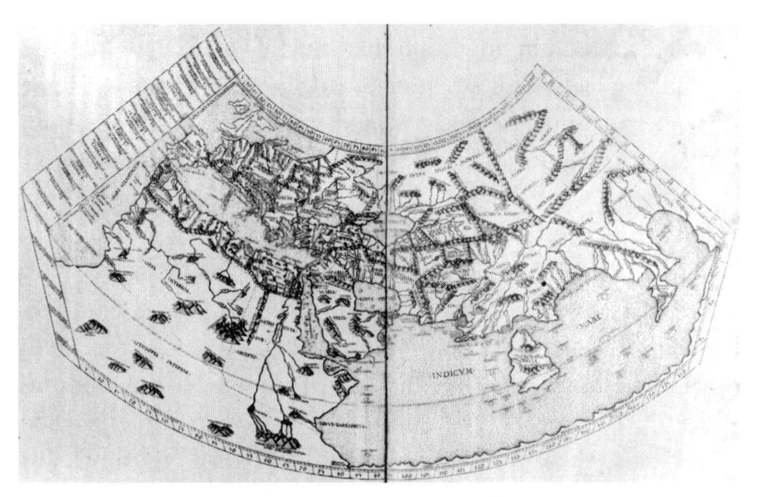

A world map from the 1400s based on the works of Ptolemy. Source: Prints & Photographs Division, Library of Congress, LC-USZ62-110342

time often emphasized a symmetrical creation, indicating God's "perfect design." On what is known as the Psalter Map (thirteenth century), Jerusalem was depicted as being the center of the world, and instead of geographical information, the map printed legends and myths depicting Christ's power overseeing the world. The map shows the world with an encircling sea and only three important waterways. The map divides the land into three continents with Asia at the top, Africa in the bottom right, and Europe in the bottom left quarter.

During the Age of Discovery (the fifteenth through the seventeenth centuries), explorers began venturing farther afield and maps became increasingly important. Prince Henry the Navigator (see chapter 2) turned the tide on mapmaking when he focused on making maps more realistic. He rejected the symmetrical and theological aspects demonstrated by the maps that were being made at that time. Cartographers copied earlier maps and then adapted them based on reports from the explorers who returned. The use of the compass, the telescope, and the sextant also improved accuracy.

The Mercator map (see sidebar) was the next big advance in mapmaking. It became the standard map projection because of its ability to represent lines of constant true bearing as straight-line segments.

When European exploration intensified in the fifteenth and sixteenth centuries, and before they had figured out how to create a Mercator projection that could be used at sea, the ship captains had Mediterranean *portolanos* (charts) showing the bodies of water, landmasses, and ports. The charts were not drawn to a grid system of degrees but were based on compass findings and estimated distances. Early cartographers interpreted the world as best they could, and some of it was speculative rationalization of what they expected it should be.

Vast portions of the western part of the United States remained unexplored and unmapped as late as the mid-nineteenth century. One bewildered forty-niner threw up his hands, tossed away his maps, and decided to "trust in good luck and when we arrive in California we shall probably know it."

The Mercator Map

Gerard Mercator, a Flemish cartographer, was instrumental in the development of maps. In 1569 he created a world map that is the essence of what is still sometimes used today because of its ability to present accurate direction and shapes on a flat surface. While the breakthrough for this mapmaking occurred in the sixteenth century, it was a good number of years before good maps could be created for nautical navigation because during Mercator's time it was impossible to determine longitude at sea. After the marine chronometer was invented and spatial adjustments could be made, the Mercator projection could be used.

Gerard Mercator. Source: Prints & Photographs Division, Library of Congress, LC-USZ62-110344

While the direction and shapes are accurate on a Mercator map, this form of projection distorts size; landmasses farthest from the equator are the most distorted, appearing larger than they actually are. Though the Mercator projection is still frequently used (including by Google maps), the problem of distortion has never been solved. On a Mercator map, Greenland appears to be almost as big as Africa, although Africa actually is fourteen times the size of Greenland. And Alaska is presented as being bigger than Brazil, while Brazil's area is actually five times that of our northernmost state. When viewed from north to south, Finland appears to be longer than India, and that, too, is an erroneous impression. As a result, the Mercator projection is still commonly used for areas near the equator, but world atlases are beginning to use other types of projections for landmasses that are farther from the equator. (A latitude of more than 70 degrees north or south means that the Mercator projection is not the best choice.)

Different Maps for Different Purposes

Every map serves some purpose, and a good one tells its story clearly so that the information is readily accessible. In mapmaking, there is general cartography and thematic cartography. While a general map can range from a world globe to a U.S. atlas, thematic maps are created for a very specific purpose. Today, maps can provide information about just about anything: the location of tanning salons, the best paths to use to hike through a national park, the likelihood of a major snowstorm in a specific area. Here are some of the categories of maps that are used regularly:

- *Road maps* are probably the most commonly used maps—certainly they are the ones we look at most often on the Internet. They show major and usually minor highways and roads, airports, railroad tracks, cities, and other points of interest in an area. People use road maps to plan trips and for driving directions.
- *Climate maps* provide general information about the climate and precipitation (rain and snow) of a region. Cartographers, or mapmakers, use colors to show different climate or precipitation zones. These are what we see when the television weather people point out the weather fore-

cast, or when we look on the Internet to check the likelihood of airline delays in a particular area where we're traveling. (My own ability to determine this is very poor. I can see an indication that there are major thunderstorms moving into the Dallas–Fort Worth area, only to board a plane that flies comfortably and bump-free into the DFW airport. Perhaps the lesson here is that experts can better interpret some of these maps than can laypeople!)

- *Physical maps* indicate the physical features of an area, such as the mountains, rivers, and lakes. Colors are used to show relief—differences in land elevations. Green is typically used for lower elevations, and orange or brown indicates higher elevations. (Leonardo da Vinci was the first to add color to a map to clarify the geography he was depicting.)
- *Topographic maps* use contour lines to show the shape and elevation of an area. Lines that are close together indicate steep terrain, and lines that are far apart indicate flat terrain.
- *Political maps*, which we see a lot of during any election year, do not show physical features. Instead, they indicate state and national boundaries and capital and major cities. During a presidential election, the United States is often depicted by a map that represents electoral votes.
- *Economic or resource maps* feature the type of natural resources or economic activity that dominates an area. Cartographers use symbols to show the locations of natural resources or economic activities. For example, oranges on a map of Florida tell you that oranges are grown there.

Purposeful Distortion

While certain maps must stress accurate distances and road placements, others are purposely amended to be functional for the information being conveyed. When Harry Beck developed a new design for the London Tube map in 1933, it was notable for abandoning scale and geographic accuracy to deliver functionality. All the stops appeared to be about equidistant, and there was no overlay of the city above. However, it couldn't

have been better for showing a subway rider what line to take to a certain stop and what stops to expect in between.

Some maps were "distorted" for commercial purposes. A map produced by the Chicago, Burlington, and Quincy Railroad Company in 1906 showed only the Chicago, Burlington, and Quincy Railroad. The routes of competing railroads were either eliminated or shown in thin, spindly lines. While travelers might have liked to know of the competing lines, the railroad had no need to send passengers elsewhere.

Jane Addams created maps of Chicago blocks that were color-coded for ethnicity and income. Englishman John Snow's scary 1855 map showed the deaths from cholera in a London neighborhood, and Florence Nightingale noted in map form the "Diagram of the Causes of Mortality in the Army in the East (1857)."

How to Read a Map

Old maps always featured a "cartouche," which is generally a highly decorated framed area of the map where the title to the map and the names of the mapmaker and the author of the map are noted. Some new maps maintain a cartouche as part of the style.

Maps generally have a "compass rose" or a "wind rose" to show the four cardinal directions: north, east, south, and west, and today maps are almost always oriented with "north" at the top. Maps contain a legend that explains the pictorial language of the map, known as its symbology.

Distance on a map can be determined by scale, which is shown by one or more ruled lines that mark off miles or other lengths of measure. On early maps, the scale indicator is sometimes decorated with a pair of dividers, an instrument used to measure distance on maps and charts.

Most maps have a grid of latitude, which measures distance in increasing degrees north and south of the equator, and longitude, which measures distance in increasing degrees east and west of a prime meridian. Greenwich, England, is often designated the prime meridian on a map. The numbers of this grid, indicating the map's location on the earth's surface, appear at the border of the map.

Personal Maps Now Possible via the Internet

Internet tools are permitting people to become cartographers, adding details to digital maps and annotating them with text, images, sound, and videos. As a result, the Internet is reshaping the world of mapmaking. Communities are sometimes posting maps of crime statistics, or schools and their rankings. You'll also find three-dimensional maps of some areas, maps noting biodiesel fueling stations in a particular area, a map of yarn stores in Massachusetts, or recent detours that recommend new traffic patterns because of construction. Much of the information is being added by ordinary people who decide their neighborhood really needs an annotated map with the best pizza shops in the area or where to go for bathroom stops in a big city.

Some of the photo services now also generate maps that show exactly where a photo was taken or use the locations as an organizing method for assembling the photographs. (For these services to work, you need to be within range of a wireless network so that the photo can be "geotagged" at the time it is taken.)

How the GPS Navigator in Your Car Knows Where You Are

Your GPS navigator—whether in your car or strapped to your wrist for a hike—works by connecting you with four satellites, each of which is equipped with a pair of atomic clocks. To determine your precise location, your GPS device constantly asks two key questions of each satellite:

- Where are you?
- What time is it?

The receiver then uses that information to calculate its own (and your) position. Microprocessors in the receiver take the four signals and calculate the position using standard geometrical methods. Satellites travel in well-defined and precisely monitored orbits, so if the orbit is known (and satellites are

constantly being monitored from base stations on the earth) and the amount of time it took your signal to reach the satellite is provided, then your location can be determined.

Three satellites would be adequate for establishing a location, but the fourth is added to increase accuracy, since personal GPS devices lack atomic clocks and therefore need another reading for accuracy. An error of one nanosecond—a billionth of a second—translates into an error of about a foot on the ground. The Department of Defense tracks minor perturbations in the satellites' orbits (from sources such as atmospheric drag and solar radiation) and sends corrections; errors can also arise from waves bouncing off buildings or mountains.

When totally exact measures are needed, a differential GPS is used. This is a fixed receiver with a known location, which measures its own location using GPS and calculates any deviation.

Early Systems

The first system was placed in orbit in 1960, with seven satellites orbiting at an altitude of about six hundred nautical miles. They broadcast signals to ground-based government users who could locate themselves by measuring the signals' Doppler shifts. They began to see that one satellite could provide a general location, two satellites created an arc that provided even more information, but three provided even greater accuracy. It became valuable to the navy for submarines and surface vessels, and then in 1967 was offered for civilian use on ships. In 1969 this was the way the USS *Hornet* was able to locate *Apollo 12* when it splashed down in the Pacific.

During Operation Desert Storm, the 1991 liberation of Kuwait, GPS satellites permitted navigation, maneuvering, and firing with great accuracy across a vast unmarked desert terrain, frequently subjected to blinding sandstorms, almost twenty-four hours a day. This minimized civilian casualties because the military was able to pinpoint strategic military locations.

At first the systems were restricted to military use, but in 1983 when a Korean airliner filled with civilians was shot down by Russia because it had accidentally drifted into Soviet airspace, interest grew in expanding the availability of these devices. At this point President Ronald Reagan announced that Navstar's signals (the name of the satellite system) would be made available for international civilian use as soon as it could be expanded.

Airlines began using the GPS, and then in September 2005, the air force upgraded the GPS constellation; this has permitted a great expansion of the system, with two separate channels now available for civilian use and great signals. Now commercial applications of GPS devices permit police to track stolen vehicles, provide locations of cell phone users, calculate the movement of the earth's tectonic plates within millimeters, and assist hikers in going through the woods. More advanced uses are being explored.

The Fun of Geocaching

With the advent of GPS technology for personal use, a new and intriguing hobby has become popular. Known as "geocaching," the pursuit involves a global treasure hunt with information given according to GPS ratings. When global positioning systems were made available to the general public, an Oregon computer consultant, Dave Ulmer, hid the first "geocache." It was a black bucket containing a logbook, a pencil, and some other small items. Online he posted the coordinates (the latitude and longitude of the geocache) and challenged others to find it. Within three days two people did.

The concept quickly grew, and the term "geocaching" has evolved. Now all sorts of individuals all over the world are involved in both hiding and seeking the hidden caches. The hobby requires a handheld GPS unit and a free account at the website www.geocaching.com. There are some basic guidelines. Most caches contain a logbook that is to be signed and dated by those who visit. The caches vary in style and substance, though they are all supposed to be in transparent containers free of advertising. Because families often "geocache" together, it is recommended that anything left in a geocache be appropriate for all ages. Caches can't be buried, placed in dangerous locations, or contain anything harmful. Some involve solving a

puzzle before receiving the first clue. Geocaches are generally "hidden" along nature trails; national parks don't allow them but state parks often do.

Searches can be found by zip code so you can limit your searches to the area around your home, or you can seek out a geocache in an area where you'll be vacationing. Many geocaches will have instructions to take something from the cache and move it to a new, specified location. Another focus of those who started the program is the principle of CITO, which stands for "cache in, trash out." People are advised to bring along gloves and trash bags so that they can leave an area cleaner than it was when they got there. The website also stresses safety first, which involves going with a partner, keeping track of your surroundings, and taking enough water.

The rangers in Colorado's State Forest State Park are big fans of geocaching and feel it's a great way to experience the park. Ninety percent of those who have come to hunt had never visited the park before. As a result, the rangers began creating their own program, and they rent out GPS devices to people who want to give it a try.

The website www.geocaching.com reports that there are currently 655,115 active sites that can be searched for, including between five and six hundred caches hidden within twenty miles of New York City. One user notes that it's "kind of like a big Easter egg hunt," and he points out that it's not that expensive and takes you to areas where you might not go otherwise.

If you haven't explored a map in a long time, take out an atlas or go on the Internet and just start looking. Chances are, you'll find it hard to pull away!

PART THREE

Pushing New Directions, Facing New Challenges

8

Westward Ho!

To truly learn something, it is helpful to narrow our vision. For that reason, the next two chapters are going to focus more tightly on the geography of the United States than the previous chapters have done. We all take for granted that our country spans from the East Coast to the West Coast, with Hawaii and Alaska added in, but rarely do we take time to think about how the geography determined the way the country expanded. For a long time, the Mississippi River was as good a dividing line as an ocean—partly because it was hard to get across but also because no one really understood what was on the other side. In this chapter we'll take a look at the westward push, and the following chapter will examine the "who" and the "what did they think?" of those who explored the most notable geographical area in the continental United States—the Grand Canyon.

But first, we need to take a step back to discuss a "western" barrier of a few hundred years ago: the Cumberland Gap.

An Early Push West: Daniel Boone and the Cumberland Gap

Myths surround Daniel Boone, the man we think of as blazing a trail, wearing a coonskin cap, and of course, "shooting a b'ar when he was only three." However, few people realize that Daniel Boone was actually key to a very early push west—when "west" just meant "a little west of east." When the country was just being settled—around the time of the Revolutionary War and just after—the Appalachians were as big an obstacle to settlers as the Rocky Mountains came to be later

on. Though we've all heard the name "Daniel Boone," few realize that one of his most notable accomplishments was to help blaze a path through the Appalachians in the area where the current-day states of Kentucky, Tennessee, and Virginia meet—what became known as the Cumberland Gap. The gap had been formed naturally by an ancient creek that flowed southward and cut through the land being pushed up to form the mountains. As the land rose even more, the creek reversed direction, flowing into the Cumberland River to the north. The gap became a trail of sorts, as it was used as a cut-through by Native Americans and migrating animal herds.

Daniel Boone (1734–1820) is a fine example of the type of trailblazers who first helped our country conquer some of the geography. He was a pioneer who was born near present-day Reading, Pennsylvania, and he spent most of his life exploring the American frontier. In 1769 a trader, John Findley, asked his friend Boone to help him find an overland route to Kentucky. That summer, Boone, Findley, and five other men traveled along the trails and made their way through the Cumberland Gap area. In 1775, when the Transylvania Company, a group of land speculators based in the colony of Transylvania (what is now Kentucky), wanted to attract more settlers to their area, they needed clear passage for settlers through the gap, and they hired Boone and thirty woodsmen to improve the trails between the Carolinas and the West. Once the pathway was cleared somewhat, Boone moved his family to the vicinity and built a village called Boonesborough in Kentucky. The trail, which ran from eastern Virginia into the interior of Kentucky and beyond, was widened in the 1790s to accommodate wagon traffic and became known as the Wilderness Road. It was to become the main route to the region (then known as the West), and it is estimated that between 200,000 and 300,000 immigrants passed through the gap on their way into Kentucky and the Ohio Valley before 1810.

That part of our country held aspects of the "wild West" during Boone's time. The British were using Indians in the area to fight for the cause of British domination, so the natives were particularly warlike. In 1776 Boone's daughter and two other girls were kidnapped by Shawnee warriors, but Boone chased

How Boone Became Mythologized

Daniel Boone became a legend early in his life when an account of his adventures, *The Adventures of Colonel Daniel Boon*, was written by a fellow named John Filson. The book was published in 1784 and made Boone well known in both America and Europe. The legend of Daniel Boone spread in 1823 when the romantic poet Lord Byron (1788–1824) devoted several stanzas of a poem to him. In the age of "manifest destiny," he was celebrated as a pathfinder who tamed the wilderness. In 1852, one writer described him as "the Columbus of the woods"; another called him "the founding father of westward expansion."

Daniel Boone. Source: Prints & Photographs Division, Library of Congress, LC-USZ62-37338

Boone acknowledged the attention with an apt remark: "Many heroic actions and chivalrous adventures are related of me that exist only in the regions of fancy. With me the world has taken great liberties, and yet I have been but a common man."

His name has long been associated with the outdoors. In 1887, Theodore Roosevelt formed a conservation organization, which he called the Boone and Crockett Club. The Sons of Daniel Boone was another organization that was to eventually become an organization that is well known today: the Boy Scouts of America.

---

after them and rescued them two days later. Two years after that experience, Boone himself was captured. He escaped and was able to warn Boonesborough residents in advance of an upcoming attack for the British that the Indians were planning.

Despite his trailblazing and heroism, Boone hit on bad times. After Kentucky was admitted to the Union as the fifteenth state in 1792, Boone lost his land because he held no clear title to it. He followed his son to Missouri and was given land in St. Charles County, Missouri, in return for his services as magistrate, but when the land was transferred to the United States, Boone once again lost his land. He spent his remaining years living with his son in the St. Charles area. As for what happened to the Wilderness Road, the area where the trail was featured the only highway through the gap until 1996. When the Cumberland Gap Tunnel was completed in the late 1990s, highway traffic was routed underground, and the historic trail was restored.

The Spanish Explore the West from the South

The Spanish took control of most of the area in the Southwest—from California to the Gulf of Mexico, including the southern Rocky Mountains and much of what is now Texas—during the nineteenth century. The two oldest cities in the United States—St. Augustine, Florida (1565), and Santa Fe, New Mexico (1609)—were the chief result of almost a century of exploration by the Spanish. The first church in North America was constructed in 1598 by the Spanish at San Juan Pueblo, thirty miles north of Santa Fe.

Spanish ruler Francisco Vásquez de Coronado (1510–1554) was the first European to explore the Southwest; he was also the first European to lay eyes on the Grand Canyon. As governor of a western province of Mexico, Coronado came north to search for the fabled Seven Golden Cities of Cibola. Coronado's exploration began in 1540. His orders were to conquer the Indians and claim the riches. For the next two years, he explored the North American continent and traveled through what is now New Mexico, Texas, Oklahoma, and Kansas, but soon he decided that the Seven Cities were nothing but a myth. Though he didn't find the wealth he was looking for, the trip was of epic proportions in opening the Southwest to further explorations by the Spanish. In a little more than two years, Coronado and his men explored much of the southwestern United States, ventured into the plains area of Kansas, descended the walls of the Grand Canyon, and visited many major Indian villages in the region. Coronado was intent on trying to get Native Americans to convert to Christianity, but that process didn't always go well; he did not hesitate to kill those whom he couldn't convert.

Why Wasn't the Santa Fe Trail as Popular as the Oregon Trail?

The Santa Fe Trail offered a southern route west, but it was problematic. After the United States acquired the Southwest during the Mexican-American war, this southern route offered a route for Southwest development. The eastern end of the trail was in Franklin, Missouri; it took travelers to Santa Fe and the Rio Grande, where they could then turn south to Mexico. Until the introduction of the railroad in 1880, the nine-hundred-mile trail was used as a vital connection between the East and the Southwest. But it was not an easy path. In addition to the arid territory, making food and water scarce, the Native Americans in the area were not friendly, and attacks on travelers were common. Anyone who lives in the West knows, too, that the weather would have presented special obstacles. Then as now, huge lightning storms could crop up with little warning, spooking the pack animals and leaving the travelers wishing for cover (of which there was none). Also unique to the area were rattlesnakes, and these were also a problem; people often died from snakebites.

In the Beginning, It Was Lewis and Clark

The story of Lewis and Clark and their groundbreaking exploration of the land west of the Mississippi is well known, but there are some aspects of it that bear retelling because these parts of the story are less well recognized. As every student has read, the Louisiana Purchase almost doubled the area of the United States. This new acquisition had an area of 828,000 square miles (2,155,500 sq km), stretching from the Mississippi River to the Rocky Mountains and from the Gulf of Mexico to Canada. Though the final agreement for the Louisiana Purchase was not completed by Thomas Jefferson until May of 1893, Jefferson sent a covert message to Congress five months before the deal was done, requesting $2,500 to fund an expedition to explore, survey, and map the lands involved in the purchase, and to meet the Indians that inhabited it.

To lead the expedition, Jefferson turned to his friend and neighbor, army captain Meriwether Lewis (1774–1809), who was serving Jefferson as his secretary-aide. Captain Lewis chose frontiersman William Clark (1770–1838) to be the coleader with him, and they made plans for the exploration to be done by the Corps of Discovery, as the group became known. Just as the sea explorers before them, Jefferson aspired for Lewis and Clark to search for an easy water route to the Pacific—a route that would be inviting to traders and emigrants alike. The route they were to follow was one determined by Thomas Jefferson, a fellow who, though he had many talents and many fields of expertise, had never traveled nor spent time in the West.

In May of 1804 Lewis and Clark left St. Louis with twenty-nine men in two flat-bottomed boats. (As they traveled farther west, they added Sacagawea, the Native American woman who made the trip with them, her husband Toussaint Charbonneau, who served as interpreter, and their infant son.)

Much has been written about their travels across the West, but for our purposes it is notable that when they arrived on the Pacific Coast on December 5, 1805, William Clark wrote: "We now discover that we have found the most practicable and navigable passage [meaning nonwater route] across the continent

of North America." But this actually was not true then, nor were they able to find a more acceptable route as they turned around to go back to Missouri. The corps agreed to divide up for the trip going home; Lewis further explored the Marias River while Clark explored the Yellowstone River to the Missouri River, where they agreed to meet before returning to St. Louis (September 23, 1806).

The expedition was a remarkable success in many ways. They brought back newly created maps and discovered 178 new plant species and 122 new animal species and subspecies (including the never-before-reported grizzly bear, the California condor, the coyote, and the gray wolf). A grizzly cub was kept on the lawn of the White House for a time, so pleased was President Jefferson with all they had brought back.

Yet they had not accomplished one of the central missions stated by Thomas Jefferson, which was to find "the most direct and practicable water communication across the continent for the purposes of commerce." The expedition had cut through the Rockies at what is known as Lolo Pass, Idaho (then referred to as the Lolo Trail). They were shocked not to find a water passage that took them through the Continental Divide and to the Pacific Ocean. Located on the border between what is now Montana and Idaho, the trail is where Lewis and Clark crossed the summit of the Bitterroot Mountains. No wagon train could follow them on their route west—the terrain was too rugged. So Jefferson's dream of opening up the West had to wait.

Though both became celebrities in their own sphere, Lewis and Clark followed very separate life paths after the journey. Clark went off to become supervisor of Indian affairs for the Louisiana Territory (1807–1821) and eventually the governor of the Missouri Territory (1813–1821), but Lewis took an appointment as governor of the Upper Louisiana Territory and began drinking heavily. When James Madison became president in 1809, he refused to pay Lewis's expense vouchers, and Lewis fell into great debt. He was on his way to Washington to plead his case for the expenses being legitimate when he stopped off at an inn about seventy miles southeast of Nashville, Tennessee. He was later found dead from two pistol

shots—one to the chest and one to the head. Captain Clark and Thomas Jefferson felt it was suicide, but another group believed he was murdered for his money and horses. Modern-day forensics could shed some light on what happened to him, but the National Park Service has been unwilling to exhume the grave of Meriwether Lewis.

Why "Westward Ho!" Took So Long

In the early nineteenth century, those in the East were not thinking "westward ho." It was very clear to Americans that the Lewis and Clark trip had been very arduous, and no one else gave them reason to think differently, though others were traversing the plains and mountains here and there. Two other early explorers were particularly discouraging in their reports:

Army lieutenant Zebulon Montgomery Pike (1779–1813) traveled throughout the area just after the Corps of Discovery had set out. In 1805–1806 he was commissioned to explore the headwaters of the Mississippi, and in 1806–1807 Pike did the same for the Arkansas and Red rivers. Pike was also supposed to identify the southwestern borders of the Louisiana Purchase. His name will always be remembered because a mountain in Colorado is named for him, though he never actually made it to the summit of the mountain. In 1806, when exploring around the Arkansas River, he came upon a peak near what is now Colorado Springs. He attempted to climb it but had to turn back because of a blizzard. Despite this, Pike's Peak bears his name. When he returned to civilization, he talked about what he had seen and reported that the plains were nothing more than a "great American desert." The thought of having to travel through dunes with no water kept other adventurers from following.

Major Stephen Long led an expedition west in 1819, and Long and his men passed through what is now Oklahoma, Nebraska, Colorado, and Kansas. He, too, sent back word that the West was unfit for human habitation. For the time being, the Native Americans were left in peace, as the white settlers stayed away thinking it was an impossible place to be.

It took another twenty years before a couple of cheerleaders for the West had an effect. John Charles Fremont, a Civil War hero and eventually a candidate for president, and his wife, Jessie, the daughter of Missouri senator Thomas Hart Benton, were destined to eventually turn the tide on the bad press the West was receiving. Fremont was particularly interested in exploring the Rockies, though he led many expeditions throughout the western territory. Fremont traveled with his wife, Jessie, along the Oregon Trail in 1842 and 1843, and Jessie wrote about what they saw. Her intelligent descriptions helped create an interest in people coming to the West.

It is also notable that it took almost 175 years for a full report on their trip to be published. The original report was a two-volume collection published in 1814; it was much abridged, leaving out most of the information about the plant and animal life observed by the Corps of Discovery. A more complete scholarly version was finally published recently. It fills eleven volumes and was published by the University of Nebraska Press from 1983 to 1997.

Business Interests Push West: John Astor and the Mountain Men

Early on, however, there were men who "smelled money" and felt the West was worth exploring. The world's richest man, John Jacob Astor, heard about Lewis and Clark's journey, and he was intrigued. Astor felt he could make money with a fur-trading operation that he thought could be based at the mouth of the Columbia River (near what is now the border of the states of Washington and Oregon). But there was a problem: He needed to devise a way to get men across the uncharted West. To accomplish this, Astor funded two different expeditions in 1810. He sent one group by ship; they were to sail around Cape Horn and reach the Columbia River by sea. The other group, led by Wilson Price Hunt, traveled overland—the first to attempt this since Lewis and Clark's trip in 1804–1806. Hunt felt that the key to getting to the West was to travel to the Snake River (starting near what is now the Wyoming-Idaho border) and travel down the river as far as they could go.

The sea-route group reached "Fort Astor" at the mouth of the Columbia, but their situation quickly deteriorated because of internal fighting. The men became angry with the captain; they destroyed the ship and many men were killed in the fight that ensued. Hunt's land-route group soon discovered that the Snake River offered no safe passage west. The river was not easily navigable, and when one of their boats upended, the group lost one of the men and a considerable amount of their supplies. The group eventually made it to the Pacific but they were starving and sick by the time they finally arrived.

The only remedy they could think of to get out of the bind they were in was to send back one of the healthier men to see if he could bring back help. They selected Robert Stuart, whose return trip took a full year—but on the way he made an incredible discovery. He found a twenty-mile-wide gap in the Rocky Mountains that was wide enough to permit wagons to come through. Located in southwestern Wyoming, the pass was located in a broad valley between two smaller mountain ranges. It simplified the journey west because it was broad and open and allowed a nearly level route between the Atlantic and Pacific watersheds. Discovery of it cut off one mountain range that had to be climbed on the northern route and extended the season for crossing, as weather was slightly less of a consideration. Eventually this became known as the South Pass; its discovery was to become the single most important one for bringing emigrants west.

To businessman John Astor, however, this discovery was "proprietary." The location of the passage west was not to be shared with other explorers, so for the next fifteen years or so it was used only by Astor's traders.

John Colter and Jim Bridger, Fur Trappers

Though few books make mention of the adventures of Jedediah Smith (see below), the American West was largely explored by these solitary fur-trapper entrepreneurs, who roamed the area even before Lewis and Clark had completed their journey. These were the unsung heroes of western exploration. They lived off the land and traveled constantly, covering thousands of miles and finding new lands as they searched for beaver pelts. Both summer and winter, they slept on the ground and ate buffalo, elk, and mountain goat, and they learned the land like no other white men before them.

In 1808 John Colter was the first white man to come upon the geysers in what is now Yellowstone Park. He described a land where hot water shot straight into the air and the earth bubbled as if boiling. No one believed him, but of course, it was true.

Jim Bridger, another mountain man, sailed down the Bear River in Utah until he found a huge body of salt water. He thought he had reached the Pacific Ocean, but he had actually found the Great Salt Lake. Bridger was well known for his trailblazing skills, and though he attempted to settle down and sell supplies to the immigrants who were starting to arrive, he soon felt it necessary to go back to a life of adventure.

Jedediah Smith, One of the Most Respected Mountain Men

Smith was thought to be the greatest of the mountain men, and as can happen, the stories of his feats grew and grew. One popular favorite was that he was attacked by an angry bear whose paw took a swipe at Smith and removed some of the man's scalp. Smith was said to have retrieved the scalp and simply sewed it back on to his own head. Smith made a major contribution to the opening of the West, however, because in 1825 he discovered the South Pass, the easier passageway through the Rockies that was being used by Astor's men.

Though the passageway had been kept secret by Astor's men, Smith made sure everyone knew about it, and it soon became part of the Oregon Trail, the California Trail, and the Mormon Trail. Approximately half a million emigrants would cross through the South Pass in the years before the railroad finally offered an even simpler method (1869). Today Wyoming Highway 28 traverses the pass. Wagon ruts that were created when it was the Oregon Trail are still visible along some parts of the highway.

Benjamin Bonneville, Undercover Agent?

U.S. Army captain Benjamin Bonneville and mountain man Joe Walker were two men who contributed to the exploration of a route to California. Walker was among the mountain men who trapped beaver and knew the West as few other men did. In 1832 Captain Bonneville hired Walker, telling everyone he was interested in learning fur trapping. Bonneville soon sent Walker to find a pass through Nevada to California, and using the Humboldt River, Walker found a way across the Sierra Nevada Mountains. The path later became known as the California Trail, and became the primary route for the immigrants to the goldfields during the Gold Rush. (When the

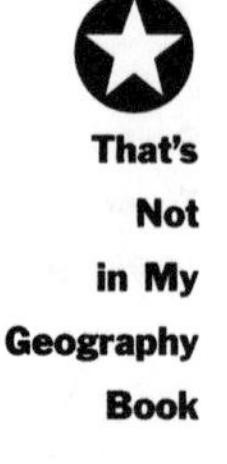

transcontinental railroad was completing its path west, its tracks were laid directly on top of Joe Walker's trail to California.)

Speculation has surrounded Bonneville's motivations for sending Walker to California. Some historians think that Bonneville was attempting to lay the groundwork for an eventual U.S. invasion of California, which was then part of Mexico.

Others Cashed In, Too

So lucrative was the fur trade that some were interested in increasing profit through better organization. A fellow named Manuel Lisa jump-started the westward movement for the mountain men, the lone fur trappers who were willing to traverse unexplored lands in search of beaver and other animal pelts. Lisa built a series of forts along the Missouri River, and he created a system where the men could travel from fort to fort to trap and sell pelts.

In 1822, William Ashley established a similar type of business but he eliminated the forts, creating the nineteenth-century equivalent of a virtual business that depended on independent contractors. Ashley ran ads in the St. Louis paper to locate trappers, provided them with supplies, and then made arrangements to meet them in a central location a year later. When they met up again, he collected the pelts, paid the men, and sent them off with fresh supplies and a new meeting date for the following year.

Ashley's business accomplished something else significant. It was the first time wagons had been taken westward, and he unwittingly began creating a wagon road for the settlers who would follow a decade later. Ashley kept the business going for fifteen or sixteen years. Over time, his men also brought him important information about the West, and those details were sent back and added to the maps being made at the time.

By the late 1830s, beaver hats were going out of style, and as a result of the diminished interest in pelts, the mountain men began to drift into other jobs. Some became farmers, while others became wagon train leaders as more and more settlers became interested in moving west.

The Oregon Trail

From the vantage point of today, it is easy to assume that the Oregon Trail was a known route west almost from the beginning, since it is talked about so commonly. But as this chapter certainly shows, the Oregon Trail was a long time in being developed; there was nothing to "discover"—it just required many feet, many hooves, and eventually wagon wheels to create the trail.

An official government survey specifies that the Oregon Trail stretched 1,930 miles—from Independence, Missouri, to Oregon City, Oregon—and the route did not even emerge until the South Pass was identified. Up until then the trip west was simply impossible for settlers. As word began to emerge—thanks to Jedediah Smith—that there was a somewhat acceptable route, a few more explorers began to give it a try.

Peter Skene Ogden, Canadian-born son of a Loyalist, was frequently sent out from Fort Vancouver to inspect other forts and note distances, and Ogden began to know and understand the West. The information he wrote down was soon sent to England and incorporated into maps, and these maps provided guidance for more explorers. Other well-known names came through the area, too. Among the scientists who came along the trail was a Harvard botanist, Thomas Nuttall, who traveled with mountain man Manuel Lisa, collecting plants along the way. Well-known John James Audubon also investigated the birds in the area.

Slowly, Oregon fever was building. The very first settlers were guided by Nathaniel Wyeth and Methodist-Episcopal missionary Jason Lee in 1834, and it was not an easy journey. The group intended to settle in Oregon Country, where they would establish a mission. At Fort Hall (built by Wyeth in southeastern Idaho), the settlers were encouraged to leave their wagons behind, venturing forth with only pack animals. While the emigrants had traveled quite far, there was still a good distance to go, making this a troubling option. Missionary Elijah White, the newly appointed Indian subagent to Oregon, led a significant expedition because they were the first to figure out a way to make the trip a little easier. With 112 emigrants following him, White figured out that they could remake their wagons into two wheeled carts, which permitted them to continue on with more of their supplies.

Though the bill didn't pass Congress until 1850, Senator Lewis Linn of Missouri first put forward the Donation Land Act as early as 1841 to extend American jurisdiction to Oregon and offer free land to white settlers and "half-breed Indians." The offer was for settlers to receive 640 free acres of Willamette Valley farmland.

Even before passage of Linn's bill, an increasing number of people were interested in making the trek. In 1843 a group led by a fellow named Marcus Whitman persevered in pushing a few of their wagons through (although Whitman encouraged them to abandon them along the Columbia River, build rafts, and float downstream to Fort Vancouver).

Getting through the Cascade Mountains continued to cause great difficulty, and until there was a solution, it was very difficult for families to make the trek. Pioneer Sam Barlow saw the need for a road when he arrived at The Dalles (the point on the trail where people had to unload their wagons and switch to river transport) in late September 1845 and noted that six families were waiting for a boat. A boat finally came for them ten days later. It was quite expensive, and they were all forced to leave behind a good portion of their supplies.

Barlow didn't want to wait so he forged ahead with wagons and considered the possibility of creating a road. He met up with another group wanting to go through, and working together the group cleared an area along an Indian path, clearing most of it by burning. When they got to the top of the ridge, an area now known as Barlow Pass, they had no idea where they were or how to proceed. Joel Palmer, the leader of the other group, climbed to the highest point that was accessible to him (probably Zigzag Glacier), observed the surrounding area, and returned to the group. Establishing some members to stay and guard the wagons at what they called "Fort Deposit," some returned along the route they had just traveled, and Barlow's group followed the Sandy River west by foot. That fall Barlow petitioned the provisional government for permission to build a road; permission was granted. Its construction allowed covered wagons to cross the Cascade Range and reach the Willamette Valley, a trip that was nearly impossible without it. Even with the road, it was still the most harrowing 100 miles (160 km) of the nearly 2,000-mile (3,200 km) Oregon Trail. Sam Barlow was quoted as saying: "God never made a mountain but what He provided a place for man to go over or around it."

Barlow had requested and been granted the right to charge a toll for two years to pay for the road: five dollars (about a week's wages) for each wagon, and ten cents for each horse, mule, ass, or horned cattle. At the end of the first season, 152 wagons, 1,300 sheep, and 1,559 mules, horses, and cattle had used Barlow Road. Though it was maintained as regularly as possible, reports of those who traveled it varied from "rough" to "barely passable." But no matter—it was preferred to the river route. During the early years, approximately three-fourths of the pioneers entering the valley came by the Barlow Road route.

In the early 1900s, the construction of the Mount Hood highway made the road obsolete. It was paved over in some sections to join present-day road systems, and in a few areas, it still exists as a dirt path.

A Geography Lesson Brought Home

Exploring the West wasn't all that was important. It was also necessary to bring people west so they could see for themselves what it was like—that you didn't have to pack a pistol to go to town and the streets weren't filled with Indians looking for a scalp. One hundred years ago—in 1908—the Democratic National Convention to nominate the party's candidates for president was held in the West for the first time. On July 8, 1908, a *Rocky Mountain News* reporter wrote:

> **For in bringing the convention to the foot of the Rockies the people of Denver have given a geography lesson to tens of thousands personally, and through the printed page to tens of millions. Those who looked upon the Missouri river as the jumping-off place of civilization have had their country's boundaries extended for several thousand miles. They realize now that there is no vacuum between the Missouri and the Pacific coast; that the whole region between belongs to the same pushing, striving, energetic race that dwells on either side.**

Like the seamen before them, the explorers who walked into the unknown parts of the United States had no maps and little reliable guidance at their disposal, and they usually had to carry with them what they needed in order to live. The opening of the West tells a great story of American bravery, inquisitiveness, ingenuity, and fortitude—all qualities that have been vital ingredients in the settling of our country.

9

The Grand Canyon

A Geographic Story

Few places in the world rival the Grand Canyon for jaw-dropping geography. This steep-sided gorge that was carved by the Colorado River (*colorado* is Spanish for "red") is known for its amazing size and beautiful landscape; people from all over the world travel to the United States to see it. The Grand Canyon is 18 miles wide and 227 miles long, with an average canyon depth of 5,000 feet (its deepest spot is 6,000 vertical feet). To travel to the bottom on foot or even with a pack animal takes a full two days. Though Cotahuasi Canyon and Colca Canyon in Peru and Hell's Canyon in Idaho are deeper, the Grand Canyon is unmatched for its breathtaking vistas, its beautifully detailed coloring, and its overwhelming size.

The Grand Canyon is primarily located in northern Arizona but it also extends into Nevada, and it is contained within Grand Canyon National Park, one of the nation's first national parks. In the Grand Canyon there are

- seventy-five different species of mammal
- fifty species of reptiles
- twenty-five species of fish
- three hundred species of birds

In 1869, John Wesley Powell was the first explorer to make a three-month river trip down the never-before-explored Green and Colorado rivers through the Grand Canyon.

How It Was Formed

Nearly 2 billion years of the earth's history have been exposed as the Colorado River and its tributaries cut their channels through layer after layer of rock while the Colorado Plateau was pushed up by long-range natural forces. Various forms of erosion contributed to the creation of the Grand Canyon.

The Grand Canyon's desert location is one of the major reasons why water has such an impact. Baked by the sun, the soil is claylike and not the type that can absorb water when it rains. (Rain in the area is also often torrential, further increasing the difficulty of the soil soaking up the moisture in an orderly manner.) The plants in the area have to be adaptive, and the result is plants with very shallow roots that are better able to grab moisture when it comes. These types of roots are incapable of retaining soil, so when it rains, the water rushes by and drains quickly into the river. The result is frequent flash floods roaring down a side canyon more like a wave of concrete than a wave of water, and with power that can move boulders the size of automobiles.

Another major factor in the creation of the Grand Canyon is erosion by ice. Especially on the North Rim, the water seeps into cracks in the rocks and freezes. This causes expansion and widens the cracks. Eventually, rocks near the rim are pushed to the edge and fall in, creating enough of a rock fall that sections of trail are frequently closed as a result.

The erosion that has occurred provides one of the most complete and viewable geologic records anywhere in the earth. John Wesley Powell described the sedimentary rock exposed in the canyon as a "great storybook," and geologists would agree. From studying the pattern of deposits and erosion, geologists are able to piece together the history of North America. Many of the formations came as deposits from warm shallow seas and swamps as the seashore repeatedly advanced and retreated over the land during prehistoric times. Some of the exposures are 2 billion years old, while other parts go back only 239 million years.

Geologists know that oceans covered much of the land area because most of the rock in the Grand Canyon is sedimentary rock, which can only be formed at the bottom of the ocean or in shallow coastal plains. The fact that it contains fossils of creatures that used to live in the ocean (brachiopods, coral, mollusks, and sea lilies, worms, and fish teeth) reinforces this belief. There was also once a tall range of mountains in the area. These mountains were eroded, over many millions of years, into a level plain. Fluctuations in climate caused the oceans to move in over successive periods, and more rock layers were deposited.

About 65 million years ago, it is thought that the Colorado Plateau began its process of uplift, and it has now risen from 5,000 to 10,000 feet (1,500 to 3,000 m). This has resulted in deep cuts in the earth, which reveal more history. As the plateau rose, it steepened the grade of the stream running through it, and as a result the Colorado River and its tributaries have increased in speed over time, which also means that

A visitor to the Grand Canyon during the early 1900s. Source: Prints & Photographs Division, Library of Congress, LC-USZ62-124385

they cut into the land faster and deeper. The terraced walls of the canyon have been created by the way different strata have been affected by the erosion.

Volcanic Activity Contributed to the Picture

Scientists also know that the Grand Canyon was not just carved by water. Volcanic eruptions have also contributed to the creation of the canyon, particularly in the western part of the canyon. New airborne elevation survey data and improved radioisotope dating of the lava in the area show that there has been an ongoing battle between water and molten rock for 725,000 years. During that period there have been at least four lava flows that halted the river in the western Grand Canyon. When the water flow is halted, the water pushes to the point that it eventually breaches the temporary dam. In other places the water may simply have eroded the rock as the river flowed over the top. Still left to ponder is how far the water backed up and what happened when the dams burst. One scientist who has been studying the dams comments on the fact that they will likely never have some answers. "It makes you wish you could have been standing on the rim of the Grand Canyon watching it all happen when those lavas were damming the river, or see when the river finally overtook the dams," notes Dr. Cassandra Fenton, a geochemist at GeoForschungsZentrum Potsdam in Germany in *Discovery News* (February 15, 2008).

There have also been small cinder cone volcanoes erupting inside the canyon. Teams of scientists have used light detection and ranging equipment (lidar, a remote sensing system used to collect topographic data) to map out the lava flows in relation to sea level, making it easier to identify the tops and bottoms of the lava flows that can be seen on the walls of the canyon.

While there is more exploration to do, scientists continue to face difficulties in navigating in and around the canyon because of the difficult terrain.

New Thinking on Time Line

Scientists originally thought that the Grand Canyon was formed by the Colorado River over a period of 6 million years, but in a recent study reported in the February 29, 2008, issue of *Science,* geologists have revised their thinking. They have now concluded that the process began more like 17 million years ago (some 11 million years earlier than previous estimates). To develop this new theory, geologists at the University of New Mexico (funded by the National Science Foundation) used an improved uranium-lead dating technique which permits dating of minerals tens of millions to hundreds of millions of years old. Using this new technology, the geologists sampled mineral deposits inside caves up and down the canyon walls, and they were able to assess the water levels over time as erosion carved out the mile-deep canyon. The geologists themselves were surprised at the age. The team was led by Victor Polyak, who was quoted in the *New York Times* (March 6, 2008) saying: "We didn't expect the canyon history to go back so far."

The National Park System Started by Teddy Roosevelt

Our twenty-sixth president, Teddy Roosevelt (1858–1919), loved the outdoors. He was both a hunter and an avid environmentalist. (In the early twentieth century, when no one realized how limited our resources were, these two interests were not mutually exclusive.) He was the first to introduce the concept that land, water, minerals, and forests needed to be held in trust by the government, not sold or given away to the highest bidder as had been the practice up until that time. He reserved lands for public use and fostered large-scale irrigation projects. As chief executive from 1901 to 1909, he signed legislation establishing five national parks: Crater Lake, Oregon; Wind Cave, South Dakota; Sullys Hill, North Dakota (later redesignated a game preserve); Mesa Verde, Colorado; and Platt, Oklahoma (now part of Chickasaw National Recreation Area). In 1908 he appointed a National Conservation Commission, with the purpose of creating the first inventory of the country's natural resources.

Another Roosevelt enactment also had a broad effect: the Antiquities Act, which was signed June 8, 1906. This act enabled Roosevelt and his successors to proclaim "historic landmarks, historic or prehistoric structures,

An early poster for Grand Canyon National Park. Source: Prints & Photographs Division, Library of Congress, LC-DIG-ppmsca-13397

and other objects of historic or scientific interest" in federal ownership as national monuments. By the end of 1906 he had proclaimed four national monuments: Devils Tower, Wyoming; El Morro, New Mexico; Montezuma Castle, Arizona; and Petrified Forest, Arizona. He also deemed the Grand Canyon a national monument, thus providing protection for it from 1908 on. Later presidents also used the Antiquities Act to proclaim national monuments—105 have been so designated. Forty-nine of them retain this designation; others have been reclassified as national parks or otherwise preserved by Congress.

Early Visitors to the Grand Canyon

The very first visitors and inhabitants in the area were Native Americans, who built settlements within the canyon and its many caves. Later the Pueblo people considered the Grand Canyon ("Ongtupqa" in the Hopi language) a holy site and made pilgrimages to it.

The first European to have viewed the Grand Canyon is thought to have been García López de Cárdenas from Spain, who arrived in 1540 under orders from Francisco Vásquez de Coronado, who ordered a search for the fabled Seven Cities of Cibola. Cárdenas arrived at the South Rim with Hopi guides and a small group of Spanish soldiers. A few of the men attempted to climb down into the canyon but had to turn back because of lack of water. The Hopi Indians likely knew a path down to the river but perhaps they didn't welcome the men. If so, their plan worked, as no Europeans came back to the area for two hundred years.

In 1776 two different groups came to the canyon area. Two Spanish priests traveled along the North Rim of the canyon in search of a route from Santa Fe to California. They eventually found a crossing at present-day Lees Ferry in northern Arizona. The other group was led by a Franciscan missionary, who attempted to convert some of the Indians to Christianity.

In the 1850s the interest in assessing the Grand Canyon grew, and a few explorers came to the area. Brigham Young sent Jacob Hamblin, a Mormon missionary, to locate a way to get across the river within the canyon. Hamblin realized the importance of befriending the native people, and over time he was able to locate accessible crossings at Lees Ferry and Pierce Ferry (both in western Arizona). In 1857 Edward Fitzgerald Beale came to the areas along the thirty-fifth parallel in search of water and a wagon road. Journals from those who accompanied him remark on the "astonishing" site before them. Also in 1857, the U.S. War Department asked Lt. Joseph Ives to attempt an upriver trip from the Gulf of California. This idea proved very difficult; Ives and his men made it to the Black Canyon area but had to abandon the plan after two months.

But the most notable early explorer was Major John Wesley Powell, who made the first recorded journey through the canyon on the Colorado River.

John Wesley Powell (1834–1902)

Anyone who has seen the rapids in the Colorado River—or seen film clips of people going through the rapids on this mighty river—will understand the enormous amount of courage it must have taken for the river to be explored. When you add in the fact that the very first exploration was led by a one-armed Civil War veteran, John Wesley Powell, it increases one's respect for his bravery and spirit of derring-do. Future generations were particularly well served by Powell, for not only was he a great explorer, but perhaps more important, he was a man of science. He had always had a major interest in geology and anthropology, and he was very aware of what needed to be observed as they navigated the river. (So rough was the going on the first trip that it took a second trip to document what he felt was important.) Eventually he was to direct governmental research into appropriate channels.

Powell was born in Mount Morris, New York, in 1834. His father was a preacher, and he moved the family west to Ohio and then on to Wisconsin, finally settling in a rural part of Illinois. Powell was intensely curious about the natural sciences. Though he studied at Oberlin College, Illinois College, and Wheaton, he never graduated. He soon began to educate himself by undertaking a good number of adventures along the Mississippi.

When the Civil War broke out, Powell enlisted in the Union army and worked as a topographer and military engineer. In the Battle of Shiloh, he lost most of one arm from a musket ball wound, but this barely slowed him down. Once his arm was somewhat healed, he returned to fight in other battles and was eventually made a major. In 1862 he married a woman named Emma Dean, and once the war ended, Powell accepted a post as professor of geology at Illinois Wesleyan University and helped found the Illinois Museum of Natural History. He refused to accept any type of permanent appointment because he wanted to remain free in order to explore the American West.

In 1867 Powell began to explore the Rocky Mountains around the Green and Colorado rivers, and he began to wonder about the canyon. Where did it go? What was it like? Could he explore there? In 1869 he combined funding from the government, the Smithsonian, and his own savings to come up with what he needed for the exploration. He selected nine men and obtained four boats and food for ten months and set out from the Green River area in Wyoming in late May of 1869. The Green River (then known as the Grand River) offered immediate challenges, with swiftly moving rapids that led to the river's juncture with the Colorado River. From there they ventured into the canyons of Utah and continued their pursuit.

There was nothing easy about the trip. Stretches with rapids appeared all too frequently. In a particularly difficult area, the men would sometimes tie the boats to a line and walk along the shore holding on to the rope attached to the boats to guide them as best they could. Occasionally they would take the boats out of the water and carry the boats and all the supplies to a calmer place on the river. But often there was no choice—turning back was not an option, and riding the rapids was the only way to continue on.

One month into the trip, one fellow, Frank Goodman, went to Powell and announced he was quitting. He said that in one month he had had "enough adventure to last a man a lifetime." (Though it is thought he found his way to safety, there is no record of what happened to Goodman later on.) Powell and the remainder of the men continued on. As time rolled by, three more men became overwhelmingly frustrated. William Dunn and two brothers went to Powell and pleaded with him to leave the river because they feared dying. They were convinced there was no safe way to continue the expedition. When Powell refused, feeling that it was actually their only way out, the three men left near the end of August. As they attempted to make their way out of the canyon on foot, they were attacked, possibly by Native Americans, and all three were killed.

Shortly after the departure of the three men, Powell and the remaining crew emerged at the mouth of the Virgin River (now under Lake Mead) and were met by settlers fishing from the riverbank. As word passed about their survival, everyone was shocked. There had been no word from them for three months, and everyone presumed they had died.

Though they accomplished what Powell had originally intended to do, he was not satisfied. While he had confirmed his theory that the river preceded the canyon and the canyon was created as the plateau rose and the water cut through the land, Powell knew that they had worked so hard to navigate that they had not adequately documented what they saw. He returned to Illinois a hero and started accepting paid lecture engagements to raise money for another expedition to photograph, document, and map where they had gone.

In 1871–1872 Powell retraced his route to take photographs and to create an accurate map and documentation of what he saw along the way. This time he employed Jacob Hamblin, the Mormon missionary familiar with the area, who had cultivated good relationships with the native people, and this permitted them to travel more safely.

While Powell's exploration and eventual documentation of the Grand Canyon were immensely valuable, he went on to contribute to geographic knowledge in many ways. From 1871 to 1879, Powell directed a federal geologic and geographic survey of western lands in the public domain and encouraged the government to initiate land-utilization projects. His report, *The Lands of the Arid Region of the U.S.* (1878), is considered a landmark in conservation literature. In 1881 Powell became the second director of the U.S. Geological Survey; he held this position until 1894, working extensively on the mapping of water sources and advancing irrigation projects. He was also the director of the Bureau of Ethnology at the Smithsonian until his death. He encouraged study and classification of the North American Indian languages and wrote on this topic. In 1895 he published a book about his explorations of the Grand Canyon. Lake Powell, a huge reservoir formed by the construction of the Glen Canyon Dam, is named for him.

Weather in the Grand Canyon

From winter snowfall in the higher elevations to a hot and arid climate in the desert locations, the weather around the Grand Canyon varies greatly—from Pacific storms bringing moisture in the winter to a late summer phenomenon known as the "monsoon season," which brings dramatic localized thunderstorms fueled by the heat of the day. The South Rim receives less than 16 inches (35 cm) of precipitation, with 60 inches (132 cm) of snow; the higher North Rim receives 27 inches (59 cm) of moisture and snowfall of 144 inches (317 cm). Temperatures vary greatly, too, from over 100 degrees Fahrenheit in the canyon to below zero along the rims.

Modern Help for a Man-Made Problem

For a very long time the weather pattern for the Grand Canyon involved heavy flooding in the spring. Snowfall from the Colorado Rockies would melt and cause massive flooding

Safety Is Still an Issue

The Grand Canyon continues to present safety hazards for those who venture there. As recently as August of 2008, heavy rains around the Grand Canyon created flooding that breached an earthen dam and placed at risk scores of campers and Native Americans who still reside in the area. The region that flooded was a side canyon containing Supai Village, where about 400 members of the Havasupai tribe live. The area is about seventy-five miles west of an area along the South Rim that is popular with tourists. Crews airlifted 170 people from the village and nearby campgrounds.

Heavy flooding from rainfall had created problems even before the dam breach. The National Weather Service reported that the area received three to six inches (7.62 to 15.24 cm) of rain Friday and Saturday and about two more inches (5.08 cm) on Sunday. Members of a boating party were stranded on a ridge at the confluence of Havasu Creek and the Colorado River and had to be rescued by helicopter, the only way to access many parts of the floor of the canyon.

during the months of May and June. This flooding used to make fast work of taking debris down the river. But between 1956 and 1966 a dam (the Glen Canyon Dam) was built at Page, Arizona, to provide water storage and to generate electricity for the southwestern states. The lake created by the dam is Lake Powell. The dam is now very controversial from an environmental standpoint.

One of the problems it has created is a significant change in the river. Natural flooding built up sandbars that are essential to native plant and fish species. The river is now cool and clear, its sediment blocked by the dam. The change helped speed the extinction of four fish species and push two others, including the endangered humpback chub, near the edge.

As a result of these changes, today's scientists realize that something needs to occur to refresh the ecosystem regularly the way it would have happened if the annual spring floods were not slowed by the Glen Canyon Dam. In the spring of 2008, for the third time in the last fifteen years (also in 1996 and 2004), scientists created a controlled man-made flood, shooting two arcs of water from the base of the Glen Canyon Dam in northern Arizona. The water level in the Grand Canyon will only rise a few feet as a result of the three-day flood, which officials hope will restore sandbars on the Colorado River downstream from the dam.

Scientists and environmentalists want to see what will happen to the fish and the canyon when the gates close at the dam and the staged flood recedes. Activists want federal officials to permanently alter the dam's operation and to adopt a seasonally adjusted plan that better mimics nature. As for how it worked, the Associated Press reported afterward that it was still unclear whether it was as beneficial as hoped.

New Plans for Mining Cause New Concerns

As the world turns to nuclear power as a possible clean source of energy, uranium has become a hot commodity for its use in the nuclear process. In the last five years, there have been 2,200 uranium claims within ten miles of the Grand Canyon's boundaries. The Grand Canyon National Park superintendent is quoted in *Wildlife Conservation* magazine for his testimony relating to his concerns about what this will do to contaminate springs and tributaries, displace wildlife, and adversely affect visitors. He was asked to rate his concern on a scale of one to ten. His concern is a ten.

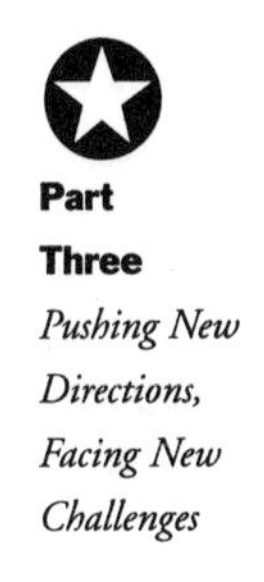

On December 20, 2007, the U.S. Forest Service approved a project being undertaken by a United Kingdom–based mining company, VANE Minerals, to conduct exploratory drilling for uranium in the Tusayan Ranger District of the Kaibab National Forest. This forest lies adjacent to the southern boundary of Grand Canyon National Park. (There were a total of five exploratory mining projects by various companies scheduled for 2008.)

There has been strong environmental opposition. Several groups sued the Forest Service, demanding that a more rigorous analysis be conducted before any more exploration takes place. In April of 2008, a federal judge issued a restraining order against VANE Minerals and Kaibab National Forest pending further proceedings. This brings to a halt uranium exploration on public lands within a few miles of Grand Canyon National Park.

Why Environmentalists Are Concerned

Another uranium-based problem exists in Moab, Utah, the former home of one of the biggest uranium-processing mills in the area. The mill closed in 1984 when the uranium boom died out, leaving 12 million tons of radioactive tailings and contaminated soil perched alongside—and slowly leaking into—the river. Unfortunately, the mining company went bankrupt in 1998, leaving the State of Utah and the U.S. Department of Energy to wrangle with safe management or disposal of the tailings in the area.

For the next few years there were government assurances that this problem would be taken care of, followed by a change of heart ("Well, maybe we won't do anything after all"), and a decision that there was a "need for another look." Then the State of Utah and the Environmental Protection Agency funded a study, and when the U.S. Geological Survey's report was released in 2008, it brought new proof that the river could wreak havoc on the pile. U.S. Representative Jim Matheson, whose district includes the Moab area, says the report was one piece of "a growing body of evidence that this was an unstable pile."

Armed with new evidence that the Colorado could shift course, undercut the tailings pile, and send the entire mess into a river that supplies drinking water for about 20 million people, Utah seems to have finally convinced the Energy Department to move the pile to higher ground. The tailings will be transported—probably by rail—to Bureau of Land Management–managed land about thirty miles north of Moab just off Interstate 70. Moving the pile to Crescent Junction will cost about $470 million, more than twice the price of stabilizing the pile and leaving it in place. And big questions remain about the final price of cleaning up the mess. The groundwater below the original site also must be treated, primarily to remove ammonia used during ore processing. The Energy Department estimates that the cleanup will take seventy-five to eighty years, but Loren Morton, with the Utah Department of Environmental Quality's Radiation Control Division, says it could actually take two hundred years—and add another $108 million to the cleanup costs.

Whether you stand on the South Rim of the Grand Canyon, hike down to the bottom with a naturalist, or look down from a transcontinental flight going over it, there is nothing like it in the world. The Grand Canyon tells a long and fascinating story of our geologic history, it provides a natural environment for people to enjoy, and it is nothing short of absolutely gorgeous. It needs to be preserved at all costs.

10

The Race to the Poles

As the world entered the twentieth century, more landmasses were settled and there was less unknown territory to explore, but the two poles at either end of the world still held great allure—unreachable, unnavigable, and therefore, all the more desirable. Explorers had ventured close enough to understand that there was little hope of future habitability in the ice-covered terrain where temperatures hovered at the freezing point on a "good" day (more typically averaging a numbing 7°F or –14°C) and some areas experiencing multiple days with gale-force winds. But interest in these areas remained high.

A recruitment ad for these expeditions might have read: "Men wanted for hazardous journey. Bitter-cold weather, long months of complete darkness, constant danger. Safe return uncertain. Small wages but honor and recognition in case of success."

None of this kept the leader-explorers from wanting to be first. The motivation of their crew of men is less certain, but like the sailors of the fifteenth century, "a job was a job"—and if there was an adventure to it, so be it.

Because exploring the area was so difficult and making progress required intelligence, ambition, and bravery, the period when most of the pole exploration occurred became known as the Heroic Age of Antarctic Exploration (ca. 1895–1917). However, not all that happened was so heroic. As a matter of fact, being able to claim to be first at both ice-covered poles was so desirous that men lied and cheated and deceived each other in order to win that race.

But before telling the stories of the races to the poles, it's important to understand the cast of characters—who the heroes of the Heroic Age of Antarctic Exploration were.

The Cast of Characters

Roald Amundsen (1872–1928) was the fourth son in a family of Norwegian shipowners and sea captains, but his mother was intent on keeping him from going to sea. She insisted he enroll in school to obtain a medical education, and Amundsen started out in the field of medical studies. He was twenty-one when his mother died, and he took the opportunity to be released of his promise to her and went off to explore the world by sea. When he was twenty-five he was part of a Belgian Antarctic Expedition (1897–1899), which gave him valuable experience at traveling under the adverse circumstances that occurred at the poles. (At this time, Antarctica—a continent the size of Europe and Australia combined—had not yet been traversed.) The group was stranded in Antarctica for the winter, an experience that provided Amundsen with lessons that he put to good use later on. Among his conclusions—based on his experience and conversations with Arctic-area dwellers in Canada—were these:

- Dogs were the most helpful "pack" animal.
- Wearing animal skins was far more effective than wearing wool parkas.
- Being certain to keep the men well fed with meat and as much fresh food as possible was key to survival.

Amundsen achieved a notable first. As early as the Age of Discovery, Europeans felt a commercial sea route from the Atlantic to the Pacific could be achieved. On a three-year expedition (begun in 1903) aboard a seventy-foot fishing boat that threaded between the Arctic islands and the Canadian mainland, Amundsen finally completed this trip in 1906. The trip was treacherous because of the ice floes, and it did not prove to be a helpful commercial sailing route because it took an inordinate amount of time. The route was not traveled in a single season for the first time until 1944, by a Royal Canadian Mounted Police schooner.

Frederick Cook (1865–1940) was an American physician and explorer who stirred up controversy in many areas of his life. Cook was tantalized by the excitement of exploration, and he served as surgeon on Robert Peary's first Arctic expedition (1891–1892). Later Cook and Amundsen served together on the Belgian Antarctic Expedition (1897–1899), and Amundsen always credited Cook with keeping the men on that expedition alive. As Amundsen told it, Cook was the fellow who realized that in order to survive, the men needed to keep eating fresh food. Cook took responsibility for hunting and providing for the men, and as a result, they spent two years in the Antarctic and were able to survive.

There is a great deal more about Cook's story later in this chapter, but it's important to know that Cook's reputation suffered from being caught in untruths about some other adventures in which he took part. Early in the twentieth century Alaska's highest mountain, Mount McKinley (also known as Denali, the official name given it by Alaska), had never been climbed. Cook attempted the climb first in 1903, but the trip failed. In 1906 he returned and said he had climbed it—claiming that he was first. Evidence to the contrary was soon presented, and as a result of the lie, all of Cook's deeds came under much closer scrutiny. (When modern-day climbers have attempted to verify Cook's claim, they have been unable to match his photos with the actual summit.)

Robert Peary (1856–1920) was a Pennsylvania native who attended Bowdoin College in Portland, Maine. (The college takes great pride in their explorer graduate, having mounted in 2008 a yearlong exhibit to celebrate the centennial of his "attaining" the North Pole.) Peary made several explorations to the Arctic, exploring Greenland by dog sled in 1886 and 1891; three more times he returned to absorb the lessons of the people and how they coped with the unforgiving climate. He noted that the native Inuits dressed in animal skins because it meant that when they traveled they could stay warm enough without having to carry extra layers of warmth, thus lightening the load. The use of support teams and supply caches also became part of what became known as the "Peary system." He was honored by the American Geographic Society and the Royal Geographic Society of London for his mapping skills and his tenacity at traveling and reaching the north tip of

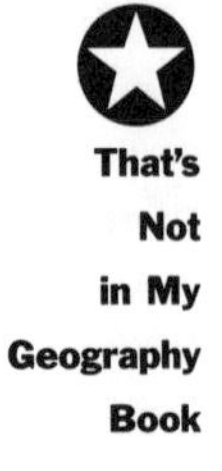

Greenland, Cape Jesup, on an expedition of 1898–1902. (The expedition was later viewed with less enthusiasm when others realized Peary had made false claims about it.)

Robert Falcon Scott (1868–1912) was a British Royal Navy officer and well-liked explorer who led two expeditions to the Antarctic. He and Roald Amundsen were the two top contenders for reaching the North Pole during an expedition during the first decade of the twentieth century.

Before he began to lead explorations, Scott served as a career naval officer in Britain. He accepted command of one of the expeditions because it offered a unique opportunity for career advancement. While love of polar exploration had not been the driving force behind Scott, he believed strongly in scientific exploration of the areas he traveled, and he was very meticulous in his observation and study.

Ernest Shackleton (1872–1922) was an explorer who grew up in Yorkshire, England. He was not much of a student in school but he read voraciously, which whetted his appetite for adventure. At sixteen he joined the merchant marine, where he achieved recognition and was invited to join Robert Scott's Antarctic expedition being planned for 1901–1902 aboard the *Discovery*. Scott's expedition ended up lasting for two summers, and Shackleton and Scott working together turned out to be a bad match. Though they made it farther south than anyone else, Shackleton was not satisfied as they were still 540 miles (869 km) from the pole. Tensions grew between the men—Scott wanted tight discipline and Shackleton would have none of it—so Scott had Shackleton sent home for "health" reasons. The two men both continued on various expeditions and were publicly friendly but were rivals all the same.

Douglas Mawson (1882–1958) was an Australian geologist and explorer who was part of the scientific staff of Sir Ernest Shackleton's Antarctic Expedition of 1907. Mawson and T. W. E. David, traveling by sledge, reached the south magnetic pole on the ice plateau of Victoria Land on January 16, 1909. (The pole shifts and has migrated some 550 miles since then.) From 1911 to 1914 Mawson led the Australian Antarctic Expedition. During this trip, Mawson also located and named the

Shackleton Ice Shelf, a sheet of floating ice bordering the Queen Mary Coast of Antarctica on the Indian Ocean. As a result of this work, Mawson was able to claim some 2,500 square miles (6,475,000 sq km) of the Antarctic continent for Australia. He was knighted in 1914 for his achievements, and from 1929 to 1931 he directed a combined group of British, Australian, and New Zealand explorers in the Antarctic.

In an Environment with No Landmarks, How Did They Know They Were at the South Pole?

Just as ocean-going vessels lost sight of any landmarks, so, too, did explorers traveling in the Arctic and Antarctica. These polar explorers were hindered by high winds, blowing snow, and cloud cover that could darken both day and night, making any type of visibility all the more difficult. Snow in the Arctic is easily windborne because it is not "sticky." At wind speeds of 60 kmh, blowing snow makes it difficult to see more than a few meters. Winter storms quickly become blizzards, and in spring and fall, landscape and clouds blend together, causing the horizon to disappear. During these arctic whiteouts, travelers can easily lose their way.

Yet time and time again, the explorers seemed certain they knew where they were going, and came back declaring they had reached the pole. With no GPS devices, how did these men determine where they had been, and more important, where they were going?

One hundred years ago—the early twentieth century—when these explorers were getting their bearings on these ice-covered landmasses, they, like the seafarers, followed some simple principles of navigation:

1. **Before they started, they determined exactly where they were.**
2. **They kept track of distances traveled using "dead reckoning" (defined as "calculation of one's position on the basis of distance run on various headings since the last precisely observed position, with as accurate allowance as possible being made for wind, currents, compass errors, etc."). This involved carefully logging compass readings, and noting distance covered and how long it took to understand travel speed.**
3. **When possible, they used sun, moon, and stars to confirm the accuracy of their dead reckoning of their position.**

A Race to the North Pole: Cook versus Peary

Between 1907 and 1909, both Frederick Cook and Robert Peary attempted to reach the North Pole. Here's how the competition got started.

Cook returned to the Arctic in 1907 but had not announced his intentions of making an attempt to reach the pole until he was already in the region. In February of 1908 he left a small community in the northern part of Greenland, taking with him two Inuit men to help out. He claimed to have reached the North Pole on April 22, 1908, but he did not return to civilization until the spring of 1909. According to Cook, they had been cut off by too much open water and had to live off local game while looking for another route home.

Robert Peary had no idea of Cook's plans. He gained funding for a trip to return to the Arctic Ocean to try to reach the North Pole in 1906, and he traveled to Ellesmere Island (above Baffin Bay, between Greenland and the islands of northern Canada) and departed from there. On April 1, 1909, Peary and five men left their base camp and by April 6 or 7 they declared they had reached the pole.

Cook returned to civilization in the spring of 1909 and made a well-received announcement that he had reached the North Pole on April 22, 1908. One week later American explorer Robert Peary returned from the North Pole and said that *he* was the first person to reach the geographic North Pole (on April 6 or 7, 1909). At first, Cook was declared the "discoverer" and Peary was referred to as the "attainer," but soon the drumbeat of bickering cropped up and has been ongoing for almost a hundred years. Peary quickly launched an attack on Cook's credibility, and while Peary fared better in the aftermath, there were still holes in his claim. Peary's records were incomplete and his diary page for the "day of attainment" was blank—until right before his book on his achievement was to be published. Cook, too, lacked documentation of his trip.

In 1989 a study of the two explorers' routes was commissioned to issue a "final word," but there is still no end to the arguing. Cook was probably an "embellisher," but did Peary make it or not? Peary is now thought to have gotten pretty

Robert Peary wearing fur from head to toe. Source: Prints & Photographs Division, Library of Congress, LC-USZ62-8234

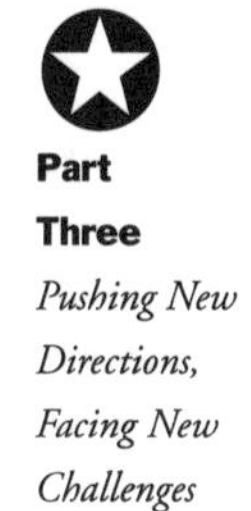

close, but debate continues since his exact whereabouts were never verified. (Later analysis indicates that Peary's measurements were not exact, and he more likely came within five miles [8 km] of the North Pole.)

As part of the 1989 study, the National Geographic Society (a major sponsor of Peary's expeditions) used "forensic photography" to analyze shadows at Peary's base camp and used technology that did not previously exist to analyze photographs

and ocean depth measures. Based on this new information, they determined that he was no more than five miles (8 km) away from the pole. They have never released the photos for independent analysis.

If anyone benefited from the ongoing battle, it was Roald Amundsen, who observed what the two explorers went through and took great precautions in verifying his trip to the South Pole. (As you'll see later in the chapter, this was critical to making his claim of being first to the pole.)

If Peary did not make it to the North Pole, then who did? That would have been intrepid explorer Roald Amundsen, who was the first person to reach the South Pole (1911); he flew over the North Pole in 1926. The first person actually to stand at the North Pole was Joseph Fletcher, who approached by air but managed to land a U.S. Air Force plane in 1952.

Human "Curiosities"

The field of anthropology—the study of man, his biological characteristics, his beliefs, and his work—was just getting

Business Is Business

"Business interests" and "crowing rights" seem hopelessly entangled in what has transpired over the past hundred years. To begin with, it is important to realize that information about these expeditions was selective. Matthew Henson, who accompanied Peary on his expeditions, and Peary both fathered children with Inuit women to whom they were not married. Peary's companions remained mum about these dalliances because they were concerned that Peary would lose backing of the voyages if the truth were known. One of the financial backers was Morris Ketchum Jesup, who was instrumental in founding an organization called the New York Society for the Suppression of Vice. Out-of-wedlock children would not have gone over well there. Finally in the 1960s, Peary's son was acknowledged.

So bearing in mind that certain truths are bad for business, we must factor in that no sooner had Peary returned from the Arctic than he learned of Cook's claim. His expedition had cost $75,000, and he had a $4,000 book advance riding on his accomplishment. Perhaps this partially explains the vehemence of his claim.

under way in the early part of the twentieth century. One of the founders, Franz Boas, who worked at the Museum of Natural History in New York City, suggested to Peary that he should bring back from one of his trips an Eskimo to study. When Peary returned in 1897, he brought back six—three men, two women, and one boy. Before delivering them to the museum, he put them on display on his ship and charged people admission to come and see them. He then delivered them to the museum, where they were interviewed, examined, measured, and put on display again for tourists to see. They actually were housed in the museum basement for a time, but in October of 1897, the *New York Times* noted that the "unfortunate little savages" had caught cold. That February, Kushan (or Qisuk), one of the men and father of the young boy, died at Bellevue Hospital. The *New York Times* (February 19, 1898) wrote: "Kushan died in consequence of the erratic weather and premature Spring. When he arrived here from Greenland last Summer with five more of his tribe, he was wearing his Summer furs, but they were too heavy for this climate, and he was compelled to don civilized costume. . . . Then he caught a

Robert Peary with Eskimos aboard his ship. Source: Marie Peary Stafford, Prints & Photographs Division, Library of Congress, LC-USZC4-7505

cold." The others got sick, too, but recovered; Kushan suffered a relapse. The article also discusses the building superintendent's intention to retrieve the body from the hospital and mount it on display. Whether or not this happened is unclear, but his son, Minik, made news later by demanding his father's body be buried, supposedly having seen it on display.

The story moves ahead to the end of the twentieth century. In 1993 Ian Tattersall, the director of anthropology at the museum, ended his ten-year search for Greenland Eskimos willing to take the skeletons of the people and bury them. Minik was the only one who survived. He was raised by the building superintendent, went back to Greenland on one of Peary's trips, and then came back to New York after realizing he didn't fit in with his native people any longer. He eventually moved to New Hampshire and became a loner, working as a lumberjack. At the age of twenty-eight or twenty-nine he died of the flu. His life story is an unremitting tragedy.

The South Pole: Another Pole, Another Race

While a few sealers and whalers may have reached Antarctica before the late nineteenth century and explorer Captain Cook had attempted to locate it as early as 1772–1774, the American flag placed at the North Pole by Robert Peary got the competitive juices of the countries going again. England felt that they needed to plant the Union Jack at the South Pole, and they raised the funds to send Robert Scott on his way. Everyone knew this would not be an easy journey. More than 99 percent of Antarctica is covered by ice, there is no indigenous population, and so bitter is the climate that there are no life-forms at all except around the coast.

Scott had been the first to explore Antarctica extensively by land. He had brought sled dogs but didn't know how to use them, so the men ended up pulling the sledges themselves. For this trip Scott brought motorized sledges, and ponies instead of dogs, so Scott seemed to have employed experience to create a better plan.

Ernest Shackleton, a well-respected leader, was on his way, too. He had to turn back, but not before attracting the interest of Norwegian Roald Amundsen. Amundsen's ship had departed from Norway, bound for the North Pole. But then he heard that Peary and Cook had already reached the North Pole, so Amundsen stopped at the Madeira Islands and took on more fresh food. When the crew reboarded after some free time, he called everyone together and announced a change in direction. They were sailing south.

More than a year after his departure from Norway in August 1910, Amundsen was in position at the Bay of Whales on Antarctica's Ross Ice Shelf. He knew that a long period of preparation would be necessary before they ventured out. His party at that point included four men and fifty-two dogs equipped with four sledges. (Among the men he had selected was one fellow who was an expert at lightening the load of the sledges.) He took along with him far more dogs than he needed, aware that some would have to be shot and consumed as fresh meat if his men were to survive. He also made several preliminary trips, stashing food and supplies along the way so that they didn't need to travel with everything they would need. Starting on August 24, Amundsen had had his men prepare for the trip each day, but upon evaluating the conditions they unpacked and stayed put for two long months, finally departing in mid-October.

To navigate through a landscape with no features, the Amundsen crew employed a specific system that Amundsen developed: They placed one of the five men on skis in the lead position to point the way and encourage the dogs pulling the sledges. The next fellow was one of the sled drivers, whose job it was to watch the compass and guide the lead skier if he strayed too much in one direction or the other from the intended route. Amundsen wrote in his book, *The South Pole*: "Imagine an immense plain that you have to cross in thick fog; it is dead calm, and the snow lies evenly, without drifts. What would you do?" Amundsen went on to explain that the second fellow, usually Helmer Hanssen, would correct the skier's direction: "A little to the right," he would shout.

Amundsen added a modification to his system: He screwed odometers to the sledges, which added a wheel measurement. By counting the number of revolutions of the wheel (and

knowing the circumference of the odometer's specific wheel), the odometer could compute the distance traveled.

In December 1911, Roald Amundsen raised the flag of Norway at the South Pole. Each man who was with him put his hand on the flag, and ninety-nine days and 1,860 miles after their departure, they were successful.

The first few days they were at the South Pole, Amundsen wrote, a gray haze had followed a fog and they could see no horizon. On December 7, for the first time in three days, they could see for a few miles around and were able to use a sextant

This photo of Roald Amundsen's crew hunting for seals was taken during their expedition to the South Pole. Source: Prints & Photographs Division, Library of Congress, LC-USZ62-70484

(see chapter 5) to verify where they were. To their own amazement, they were exactly where their dead reckoning estimated them to be (88 degrees 16 minutes south—a minute is 101 feet or 31 m). To be at the pole, they had to reach 90 degrees south, and on December 14 at 3 p.m. the drivers who were watching the compasses and sledge meters called out "Halt!" Amundsen wrote that even they knew they could not be *exact*, given the type of equipment they had, so they marked their best guess and then hedged their bets by circling the area around where they thought the pole would be. Three of the men, in an effort to be certain that one of them definitely walked across "the spot," walked out from camp in three different directions for twelve miles in each direction. They took additional measurements to further document their positioning. On December 16 and 17 they took hourly observations of the sun using the two sextants they had with them in order to further verify their positioning. They decided they needed to adjust their position a bit, so they walked another four miles south before turning around to travel back to their ship.

As for Scott, they were experiencing many problems. A good number of the men turned back, leaving Scott and four companions to travel on their own. They reached the pole on January 17–18, 1912, but were disheartened to see that Amundsen had arrived first. As they returned, they ran into terrible weather. They were short on food, and at the end of March 1912, those who remained died asleep in the tent. A year later (November 1912), a search party was sent out and located the camp and the bodies. The bodies were found but left and the writings were recovered so people know what happened. Scott and his party were only eleven miles away from a food depot they had set up earlier.

What Happened to Shackleton, Whose Name Lives On as an Exemplary Leader?

Even though Amundsen had accomplished the shared goal of reaching the South Pole in 1911, the public fascination with the Antarctic continued. Shackleton continued to seek backing for a return trip for a trans-Antarctic expedition. Out of the

five thousand people who applied to go with him, Shackleton chose fifty-six, based on reputation and camaraderie. He expected everyone from the scientists on down to do chores, so it was important that the selection be made correctly.

One of the ships, the *Endurance*, ran into terrible difficulty on the voyage, and its hull was damaged on a giant ice floe. Shackleton ordered the men to abandon ship and for two months they camped on an ice floe. Finally Shackleton left the party to go for help. He took with him several of the stronger men and the better lifeboat, knowing that they needed to travel eight hundred miles of open ocean to find help. The men he chose were malcontents, whom Shackleton knew he could not afford to leave behind, or they would cause trouble. The weather was so bad during this "mission for help" that Shackleton and his men were sailing somewhat "blind." Celestial navigation readings were possible only four times during the trip. Once they finally reached land, Shackleton and two of the men spent thirty-six hours climbing across rugged terrain in order to reach civilization. In attempting to return to the men, there were three failed attempts, and finally Shackleton asked the Chilean government to send a tug to help. The tug finally reached the men and evacuated all twenty-two men, who had been stranded for 105 days.

Shackleton took several more trips to the Antarctic but had health problems. In 1921–1922, Shackleton chose to return yet again and wanted to circumnavigate the Antarctic. At the age of forty-seven, Shackleton suffered a fatal heart attack on this journey. Shackleton did not achieve immediate recognition but what he went through has become an oft-told story of building and managing a great team and acting heroically. The story has become the focus of many books and movies.

Shackleton's death marked the end of the Heroic Age of Antarctic Exploration, a period of discovery characterized by journeys of geographic and scientific exploration in largely unknown areas, without the benefits of modern travel or radio communication. One of Robert Scott's travel companions on the *Terra Nova* expedition, Apsley Cherry-Garrard, wrote: "For a joint scientific and geographical piece of organization, give me Scott; for a Winter Journey, Wilson; for a dash to the Pole and nothing else, Amundsen: and if I am in the devil of a hole and want to get out of it, give me Shackleton every time" (quoted by Sara Wheeler in *Cherry: A Life of Apsley Cherry-Garrard*).

Today Antarctica is shared by twenty-seven nations, all of whom have scientists based there. They study climate change, the destruction of the ozone layer, and basic preservation of the species.

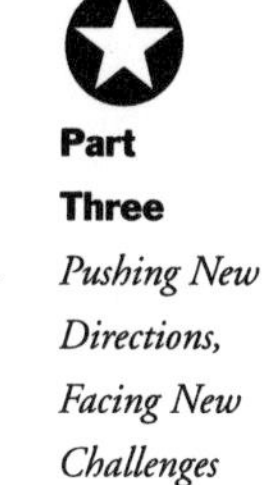

11

Who Are the Women Explorers?

When it comes to exploring, "dead white guys" are about the only people you ever read about in history or geography books. However, there were some amazing contributions by women who were also driven by a "need to know," or a "need to see what's on the other side," or a need to learn about a world other than their own.

In this chapter, you'll meet some notable women explorers and learn about their contributions to the world as we know it.

Ida Laura Pfeiffer (1797–1858): Bringing the World to Life

Just as boys dreamed of adventure, so, too, did some girls. Ida Pfeiffer was an Austrian woman who waited until she was forty-five and her children were grown in order to fulfill her dream of adventure and traveling to foreign places. Her first trip was along the Danube River to the Black Sea and Istanbul, and on to Palestine and Egypt before coming home through Italy. She wrote of her journey in *Reise einer Wienerin in das Heilige Land* (published in Vienna in 1843), and the money from book sales permitted her to continue to travel. In 1845 she wrote a two-volume book on her travels in Scandinavia and Iceland. In 1846 she set out to travel around the world, heading south so she could visit Brazil, Chile, and other countries of South America, and continuing on to Tahiti, China, India, Persia, Asia Minor, and Greece. The trip took two and a half years. In 1851 she went to South Africa with the intent of visiting the African interior. When this proved impractical, she made a second trip around the world, reaching home again in

Ida Laura Pfeiffer. Source: Prints & Photographs Division, Library of Congress, LC-USZ62-108109

1854. This book was published two years later: *Meine zweite Weltreise.*

Pfeiffer was gifted at bringing her travels to life so that the armchair traveler had the feel of the places she visited, the people she saw, the culture they lived, and the weather she experienced on the way. She also brought back specimens of plants from her travels. When in China, she wrote of footbinding, something that might have been perceived differently by men: "Four of the toes were bent under the sole of the foot, to which they were firmly pressed, and with which they appeared to be grown together; the great toe was alone left in its normal state. The lower portion of the foot was scarcely four inches long, and an inch broad."

In May of 1857 she went to Madagascar. She was initially well received, but she and some other European visitors became mixed up in a plot to topple the queen and they were all expelled from the country. Pfeiffer had become ill while there and went home and died in 1858. She was a member of the geographical societies of Berlin and Paris but was kept out of the Royal Geographic Society in London due to her sex.

Geraldine Moodie (1854–1945): Arctic Photographer

Geraldine Moodie was a Canadian woman who married a fellow who became part of the Northwest Mounted Police. She was to become the first woman photographer to work above the Arctic Circle and the first to make a record of the Inuit people.

While raising six children, Moodie pursued watercolor painting, but she soon took up photography. From 1891 to 1900 she had professional studios at several locations. In 1895 her work was well enough known that the prime minister of Canada commissioned her to photograph sites in northwest Canada, where the native people had fought back against the efforts of a new government. She also took photographs of the Cree people who were native to the area. In the early twentieth century she joined her husband at his post as superintendent of Canada's Northwest Mounted Police in the eastern Arctic and the Hudson Bay district, and she photographed the Inuit people as well as the wildflowers in the area. In 1912 they moved farther north and were posted in the Yukon for several years. Her images are preserved in both the British Museum and the National Archives of Canada.

Moodie was raised in a family of successful women. Her grandmother, Susanna Moodie, had written a book, *Roughing It in the Bush*, about her experiences in the frontier. Her grandmother's sister, Catherine Parr Traill, wrote books on the flowers of Canada and was also well known for her work.

Lady Anne Blunt (1837–1917): Arabian and Middle Eastern Traveler

Anne Isabella Noel Blunt was born into nobility and was very well educated as a young girl. She was also an accomplished equestrian. Her marriage to poet Wilfrid Scawen Blunt seemed like a good match; his interest in the Middle East and her fascination with Arabian horses led them to travel to the Middle East and to bring back some of the best Arabian horses available at that time (a line known as Crabbet). Over the years they made many trips to Egypt and Arabia, buying horses from Bedouin tribesmen as well as from Egyptian royalty. She kept careful journals of her travels and two books resulted: *Bedouin Tribes of the Euphrates* and *A Pilgrimage to Nejd.*

Lady Blunt's marriage, however, was not a happy one. Together she and Wilfrid had one daughter, Judith, but unfortunately, all of her other pregnancies ended in miscarriage or the birth of a baby who died soon after. Wilfrid was a difficult man both as a husband and a horse owner. As a husband he thought nothing of having multiple mistresses; as a horse owner, he espoused the belief that the horses were from the desert so they should live under "desert conditions"—meaning that they were not given adequate food and water.

By 1906 Lady Blunt asked for a separation, and she spent more and more time—and eventually full-time—at her estate in Cairo. She continued horse breeding, a legacy that was carried on by her daughter. To this day, the vast majority of purebred Arabian horses trace their lineage to at least one Crabbet ancestor.

Fanny Bullock Workman (1859–1925): Climber, Geographer

Fanny Bullock was born and raised in Worcester, Massachusetts, and her life of travel began when she married American physician and explorer Dr. William Hunter Workman (1847–1937). But she was a geographer, cartographer, explorer, and mountain woman in her own right. Most of her work was done in the Himalayas. Workman set the climbing record for women on Nun Kun (23,300 feet), setting the women's altitude record in 1906. She and her husband operated with a true partnership, sharing responsibilities and trading on and off with specific jobs so there was no gender difference in their roles. They wrote several books together about pioneer exploration and climbing in that part of the world.

Fanny Bullock Workman. Source: Prints & Photographs Division, Library of Congress, LC-USZ62-118944

Workman was a big proponent of women's suffrage and women's rights, and she felt that many of the men she encountered—both scientists and fellow climbers—were very antagonistic to her because she was a woman.

Mary Henrietta Kingsley (1862–1900): Sheltered Daughter, Avid African Traveler

Mary Henrietta Kingsley was an outspoken critic of European colonialism, a champion for indigenous customs, and an acknowledged nineteenth-century authority on West Africa. She was the daughter of a well-respected physician who traveled widely, accompanying various well-off patients on their world travels. Dr. George Kingsley kept meticulous notes on his travels and hoped to publish his information at a later date. Her mother was an invalid, and while her brother was educated at Cambridge, Mary was expected to stay home to look after her mother. While she was not given much education, she did have access to her father's library and made good use of her time; eventually her father's health failed and he relied on Mary to work with him to help organize his notes.

When both her parents died within a few weeks of each other in 1892, Mary was expected to live with her brother. But when he set out for China in 1893, Mary was left with an income. She, too, decided to travel, but first she sought advice from those around her. All discouraged her idea, but physicians—pointing out the geographical maps of tropical diseases—mentioned that if she did go, she should bring home specimens for study. Her father had begun a book on Africa, so she decided to return there to research the book, and the native people in Angola taught her the skills she needed in order to survive. A few years later, in 1895, she returned to Africa to study cannibal tribes. She traveled in remote areas and collected specimens of fish that were thus far unknown. At Mount Cameroon, she climbed the 13,760-foot height by a route not taken by any other European. Traveling in the Congo area, she learned sailing skills from the boat captain on whose ship she sailed from England. (Her original visit was to

the area made famous by Joseph Conrad's *Heart of Darkness*, where she encountered hippos and crocodiles.)

Kingsley's first book describes hair-raising adventures in an understated way. Among the adventures she describes are being caught in a tornado when climbing Mount Cameroon and falling into a fifteen-foot game pit with poisoned stakes created to trap animals, when she was literally "saved by her petticoat." She traveled in the clothes that were customary in England, including a long black skirt with a petticoat and high waist, high collars, and a small fur cap. She later stated that the petticoat was so thick that it prevented her from being impaled on a stake.

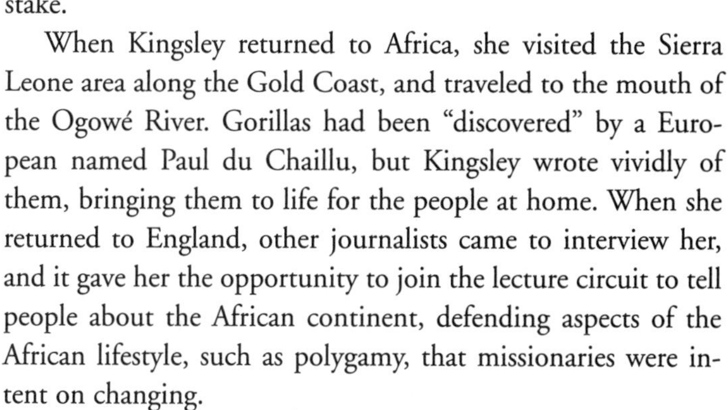

When Kingsley returned to Africa, she visited the Sierra Leone area along the Gold Coast, and traveled to the mouth of the Ogowé River. Gorillas had been "discovered" by a European named Paul du Chaillu, but Kingsley wrote vividly of them, bringing them to life for the people at home. When she returned to England, other journalists came to interview her, and it gave her the opportunity to join the lecture circuit to tell people about the African continent, defending aspects of the African lifestyle, such as polygamy, that missionaries were intent on changing.

Kingsley wrote two books about her travels, *Travels in West Africa* (1897) and *West African Studies* (1899). She died during the Second Boer War, when she contracted typhoid while treating prisoners. A Mary Kingsley Society was formed in her honor and eventually it became the Royal African Society. Though the British disagreed with Kingsley's anticolonization sentiments, they were enthralled by her stories of the native peoples.

Gertrude Bell (1868–1926): Archaeologist and Intelligence Officer

Gertrude Bell was one of the most powerful women in the British empire. She became active on the international scene after living in the Middle East for a good number of years. She explored and excavated many of the ruins in the area of Iraq

and founded the archaeological museum in Baghdad. She understood the importance of establishing relationships with the Bedouin tribes in the area. She was an archaeologist, a traveler, and a government official; she helped mold the postwar administration of Mesopotamia and helped bring Faisal to the throne of Iraq in 1921.

Bell grew up in a wealthy family in the English countryside. She was tutored at home but then decided to attend Oxford University, an unusual decision for that time. In 1892 she traveled to Persia (Iran) to visit an uncle, and this experience whetted her appetite for adventure. Over the next few years, she traveled around the world twice and achieved recognition as a mountaineer through her climbing in the Alps. In 1901 she returned to the Middle East, became a well-respected archaeologist, and documented her finds both in writing and in photographs. She excavated ruins of both Byzantine and Christian churches and spent several years searching for ancient ruins in the desert, which prepared her for her role in World War I.

When World War I broke out, she was recruited by the British to work in the intelligence unit and she took the position of Oriental secretary to the British High Commissioner. Her work was notable because it was helpful in the Mesopotamian area transition from fiefdoms to modern states. As Oriental secretary, Bell was influential in establishing Iraq as a nation with Faisal as the king. She drafted many of the laws of Iraq and was instrumental in setting laws regarding education for women.

In January 1919 when the Ottoman Empire collapsed, Bell was assigned to write an assessment of Mesopotamia and options for the future of Iraq. She later served as a liaison with the government as the British worked to create a country inhabited by a Shi'ite majority in the south, and Sunni and Kurdish minorities in the center of the country and the northern section respectively. By keeping the Sunni and Kurds part of the country, they intended to keep control of the oil fields. The Shi'ites, which had the majority, were viewed as too volatile to govern and the British thought the Sunnis should lead. As we know now, these rivalries and different religious attitudes have never settled down in Iraq.

Things did not work out much better for Bell, who returned to Britain in 1925 and was unwell. It is not known whether the overdose she took was an intentional suicide or accidental.

Annie Montague Alexander (1867–1950): Explorer, Collector

Annie Montague Alexander was born in Hawaii and was part of a wealthy family that permitted her to follow a life of adventure. Her father included her on sailing and bike trips as well as a safari. At a time when few women undertook activities outside the home, Alexander became an explorer, a naturalist, a skilled markswoman, and the founder of two natural history museums in California.

In 1900 she began attending lectures on paleontology by John C. Merriam at the University of California. She approached Merriam about setting up her own fossil-hunting expedition, and she began traveling and collecting fossils but also studying live animals. (She realized that to understand the fossils there were some species still living that could be studied.)

In 1904 her father died suddenly when they were together on a safari, and Alexander decided to make collections of West Coast fauna. She began collecting fossils and gathered a collection of tens of thousands of specimens of mammals, birds, amphibians, and plants. These became the core collection of the Museum of Vertebrate Zoology, which she founded in 1908 at University of California, Berkeley. Along with Louise Kellogg, her partner of forty years, Alexander collected thousands of animal, plant, and fossil specimens throughout western North America. Their collections serve as an invaluable record of the flora and fauna that were beginning to disappear as the West succumbed to spiraling population growth, urbanization, and agricultural development. She had fifteen fossil, plant, and mammal species named for her, including one of the largest of Alaskan bears, *Ursus alexandrae.* Several others honor Kellogg, who continued to make field trips after Alexander's death.

Alexander's dealings with scientists and her encouragement—and funding—of women to do field research earned her much admiration, even from those with whom she clashed. She was a keen observer of human nature who loved women and believed in their capabilities. Her legacy endures in the fields of zoology and paleontology and also in the lives of women who seek to follow their own star to the fullest degree possible. Her eightieth birthday was spent on a three-month collecting trip to Baja. She left a legacy in the two museums she founded and eventually endowed.

Sue Hendrickson (1949–): Deep Sea Diver and Dinosaur Hunter

Curiosity and serendipity helped Sue Hendrickson carve out an ideal life as an explorer. She was raised in Indiana but left high school before graduating. She moved to Florida and then California where she sailed and earned a living helping with boat maintenance. (She eventually picked up a high school diploma.) One day she stopped in at a tropical fish store and became fascinated with the fish and how the store owner got them. She immediately decided that diving for tropical fish was going to be her career. She moved back to the Florida Keys and began diving with a group of people who caught and sold fish.

During a dive in the Dominican Republic, a friend showed Hendrickson a fossilized insect preserved in amber. She was fascinated and began reading all she could about what was a 23-million-year-old fossil. Soon she began visiting the amber mine regularly and started providing many museums and universities with amber fossils. This work put her in touch with a paleontologist who invited her to join him on a dig for whale fossils in Peru. She worked for six winters in the deserts of Peru, where they uncovered numerous whale fossils.

But her biggest find was yet to come. In Peru in the 1980s, Hendrickson met a fellow who was digging fossils in the United States. So for three summers, she traveled to South Dakota to work at digs in that area. One day toward the end of the summer of 1990, the truck they were traveling in got a flat.

The others went off to see about getting help, and Hendrickson and her dog stayed around the area just looking around at some of the areas they had not yet had time to explore. As she examined the earth, she noted something sticking up a bit that looked like a bone. As she looked up a bit, she thought she saw what could be dinosaur backbones.

Hendrickson notified the group of her find and they went back to dig. It took five days to remove the thirty feet of rock that covered the dinosaur's skeleton. All the work had to be done by hand because machinery might have caused damage. After three weeks Sue, as the dinosaur skeleton became known, was finally fully removed from the earth, and it was taken back to the Black Hills Institute.

The first *Tyrannosaurus rex* skeleton was found in 1900. Since then only seven skeletons have been uncovered, and none of them are more than half complete. Finding Sue provided scientists with the largest, most complete, and best-preserved *T. rex* ever found. The Field Museum in Chicago bought Sue for $8 million, and Hendrickson continues to pursue her dual focus on underwater archaeology, uncovering shipwrecks and remains, and fossil hunting, spending summers in the western United States, looking for more dinosaur bones.

Like Hendrickson, all of these women explorers prove that whether you are male or female, it is perfectly possible to pursue your dreams and "be whatever you want to be." They all did.

PART FOUR

The World Depends on Us

12

New Ideas on Environmental Geography

One of the most important recent works on the topic of taking care of our world is the 2005 title *Collapse: How Societies Choose to Fail or Succeed*, written by scientist, professor, and author Jared Diamond, who is also the author of the Pulitzer Prize–winning book *Guns, Germs, and Steel.* In *Collapse*, Diamond provides a wide range of case studies that pinpoint a variety of reasons why various societies no longer exist. The problems he highlights include climate change, warring neighbors, a lack of adequate countries with which to trade for essential and necessary goods, environmental problems, and the individual society's response to environmental problems. His examples include societies that have faced complete extinction (Pitcairn Island), population crash (Easter Island), resettlement (Vikings), civil war (Rwanda), anarchy (Somalia, Haiti), or even just the demise of a political ideology (the disintegration of the Soviet Union).

He makes a strong case for the fact that top-down (government-led) as well as bottom-up (citizen-led) environmental management is going to be necessary in order to manage the issues we currently face, including destruction of natural habitats (mainly through deforestation); reduction of naturally grown food; loss of biodiversity; erosion of soil; depletion of natural resources; pollution of freshwater; maximizing of natural photosynthetic resources; introduction by humans of toxins and alien species; artificially induced climate change; and, finally, overpopulation and its impact.

We need only look around in our own homes to think of what is happening in our world. Physical possessions—including difficult-to-recycle electronics—have multiplied in recent

years, and as our society has become more "convenient," we've increased the number of things that we consider disposable. We are only now learning the error of our ways with plastic bottles and bags, for example. Not only is bottled water often simply an expensive form of tap water, but 90 percent of all plastic bottles end up in landfills, where they take thousands of years to decompose. And every year, Americans throw away some 100 billion plastic bags after they've been used to transport a prescription home from the drugstore or a quart of milk from the grocery store. It's equivalent to dumping nearly 12 million barrels of oil. Only 1 percent of plastic bags are recycled worldwide—about 2 percent in the United States —and the rest, when discarded, can persist for centuries.

What's going to happen to all the "big stuff" we acquire—the computers that seem to grow old (and break down beyond repair) within about five years, the hybrid car batteries that make better use of fuel but end up being a major headache when the car is no longer wanted since the batteries are big and simply don't decompose? Does anyone remember the garbage barge that left Long Island, New York, bound for a landfill in the South, only to find that no one wanted that garbage anymore? In 1987 that barge sailed around for a long time looking for a place to dump its contents. What will we do when our entire world is overflowing with effluence from our affluence?

Diamond's point about government-citizen partnership makes a great deal of sense. While an individual can decide to walk to the store instead of driving, it takes a government to set up watchdog agencies on manufacturing plants here and abroad to push for standards that are environmentally wise.

While Diamond brings up many environmental issues—all of which are very important—I want to close this book with one that is often underrecognized. I grew up in the arid West where "liquor was for drinking and water was for fighting over." As a result, I know what the threat of a drought can mean to a community. For those people who live in states where there seems to be plenty of rain—even some periods when flooding occurs—water seems to be so plentiful that people would never expect a true shortage. This is an inaccurate perception, so this chapter will conclude with a bit about why we need to rethink our water usage.

Water, Water Everywhere?

Many people live in a rain-filled part of the United States where a water shortage seems unthinkable—or it seems like the kind of local problem that will be handled quickly and easily after some negotiations by the city or state involved. And when we read of water feuds between states like Arizona and California, they don't seem to affect "us." That is, unless you've lived in a community where you've experienced water shortages. One community that will always listen carefully to discussions of water shortages is Beulah, Colorado, a mountain community where it was so dry in 2006 that the water supply for the mountain community totally dried up. For several weeks, the only water available to residents had to be trucked in regularly. This is a scary thought, particularly when you read more of this chapter and consider that this kind of shortage is growing increasingly likely because of climate change and tremendous growth in population (we'll soon be at 7 billion people in the world).

As it happens, many countries are already dealing with severe water shortages. The World Bank reports that eighty countries now have water shortages that threaten health and economies, while 40 percent of the world—more than 2 billion people—have no access to clean water or sanitation. More than one in six people lack access to safe drinking water, according to the World Water Council. What experts fear is that these numbers will grow, and one day there will be no peaceful solution. What if countries no longer have enough water to drink or to irrigate crops to grow food? (Think the "Dust Bowl" era in the United States.) What happens if the better part of a population has no food to eat? If wealthier countries are also overpopulated and having trouble finding enough water, airlifting food to these people is no longer an option, and at some point it will occur to someone that water is, indeed, "for fighting over."

While the Middle East may be flush with oil, water is another matter. They need to negotiate with Israel and Syria regarding access to water. In recent years, Iraq, Syria, and Turkey have exchanged verbal threats over their use of shared rivers. (Ironically, the word "rival" comes from a Latin root and means "someone who shares the same stream.") Many of the nations with tight water supplies receive most of their water from rivers that flow through bordering countries that are considered unfriendly (Botswana, Bulgaria, Cambodia, the Congo, Gambia, Sudan, and Syria, to name a few).

Water, like energy in the late 1970s, will probably become the most critical natural resource issue facing most parts of the world by the start of the next century. World population has recently reached 6 billion and United Nations projections indicate 9 billion by 2050. According to the United Nations Population Fund, global consumption of water is going to double every twenty years. What water supplies will be available for this expanding population? In 2005, the Environmental Protection Agency estimated that aging systems in the United States will require $277 billion in investments just to upgrade and maintain drinking water quality over the next twenty years. Even if the water magically appears, delivering it in a clean and acceptable state costs big money.

Like Jared Diamond's approach to environmental sustainability, water, too, needs both a top-down and a bottom-up solution. The ever-increasing world population is certainly one drain on our water supply. As populations grow, industrial, agricultural, and individual water demands escalate. According to the World Bank, worldwide demand for water is doubling every twenty-one years, more in some regions.

But a higher standard of living has also increased the water used. As diets contain less grain and more meat, it takes more water for an increase in the agriculture needs for animals that are later slaughtered. In addition, water quality is declining in many areas of the developing world as population increases and salinity caused by industrial farming and overextraction rises. About 95 percent of the world's cities still dump raw sewage into their waters.

Climate change represents a wild card in this scenario. If, in fact, climate change is occurring—and most experts now agree that it is—what effect will it have on water resources? Some experts claim climate change has the potential to worsen an already gloomy situation. With higher temperatures and more rapid melting of winter snowpacks, less water will be available to farms and cities during summer months when demand is high.

Desalination (a process whereby the salt is removed from salt water in order to make the water safe to drink) is often cited as a possible solution. Some researchers fault the United States for not pursuing this line of research as Saudi Arabia, Israel, and Japan have done. (There are approximately 11,000 desalination plants in 120 nations in the world, 60 percent of them in the Middle East.) But the process is complex and expensive; it is not a "no-brainer" solution.

Right now water is generally viewed as a free and natural commodity, and some feel that water management needs to be

Investing in Water

Anything that gains value eventually becomes something people want to invest in. Since the summer of 2007, a high number of investment vehicles have hit the market to capitalize on the rising global need for clean water ("Commoditization, anyone?"). Unlike other commodities, water is traded in regional markets, not global ones, because of its very nature—it's heavy and difficult to transport.

Most municipal water utilities are monopolies and are not publicly traded. Only a few other types of water-related companies have recently gone public. American Water Works may become the biggest publicly traded water utility in the United States, and Energy Recovery Inc., a desalination-technology firm, filed to go public in April of 2008. Perella Weinberg Partners Oasis Fund is a $500 million hedge fund focused on water, clean technologies, and alternate energy. There are also several actively managed funds that focus on the water sector, but it's tricky to capitalize on the need for water because there are so many steps between source and delivery. While investment dollars have surged, the stocks are still viewed as quite volatile.

managed internationally by a free market model. If water rights were bought and sold more commonly (as they are in the western United States), then it might heighten conservation as well as the ability to establish fair trade. (See sidebar "Investing in Water.")

Conservation—doing more with less—is also desirable. One need only visit Scottsdale, Arizona, an area surrounded by deserts but filled with green golf courses, to speculate that there might be a better use for the available water than keeping fairways green. Many communities in the West are undertaking community education about what plants grow well in arid climates. If all areas could promote greenery that fits with the climate, that would be a strong first step toward better water management by communities and individuals. Ultimately, however, an awareness of the global water crisis should help people—both governments and individuals—make wiser choices, if they are reminded that water is a very scarce and very valuable natural resource.

And here's an interesting footnote: The one good thing about an economic downturn may be a reduction in water usage. Water usage actually drops during a recession because 40 percent of fresh water in the United States is used by industry. So when business slows, use of fresh water goes down. The power industry is one of many big water users; water is critical in hydroelectric dams and in cooling processes at fossil-fuel and nuclear-power plants. The food industry also uses a tremendous amount of water. It takes 260 gallons of water to produce 2.2 pounds of wheat and 3,380 gallons of water to produce 2.2 pounds of beef, according to the World Water Council.

A Case History? Australia's Plan

Australia has always been the world's driest inhabited continent, so when the drought of 2007 hit the outback, it felt particularly harsh. For years, development in Australia relied on building reservoir systems so that parts of the interior of Australia could be used to grow crops, including the very lucrative cotton crop. As the drought worsened, Australia's government

William Mulholland (1855–1935)

William Mulholland was an Irish immigrant who arrived in New York in the 1870s and eventually moved west. He was a self-taught engineer, eventually rising to become the head of the Los Angeles Department of Water and Power.

During his tenure, he oversaw completion of construction of the Los Angeles Aqueduct, an engineering marvel that stretches more than two hundred miles through mountains and over desert to bring Los Angeles the water it needs to grow. Tapping the Owens River in the Sierra Nevada, the aqueduct transforms the once-fertile Owens Valley into a watershed for what has become one of the most populous cities in the nation. Mulholland is the perfect fellow to spotlight in this chapter because the building of this aqueduct introduced a wide variety of issues that arise every time water is redistributed in the West—or anywhere in the world, for that matter.

The water first flowed into Los Angeles in November of 1913; fifteen years later, Owens Lake (from which the water had been diverted) was totally dry. This started what became known as the California Water Wars (a fictionalized version of this story became the film *Chinatown*). As the issues became clear, a case was made that the water rights dealings had been underhanded and had unfairly been taken from the Owens Valley farmers, who retaliated by dynamiting parts of the aqueduct.

Mulholland's reputation was further eroded when in 1928 he personally inspected a dam that held back water above Santa Clarita Valley, north of Los Angeles. Hours after having been given the okay by Mulholland, the dam gave way and 12.5 billion gallons of water flooded into the Santa Clara area toward Ventura. The town of Santa Paula was buried under mud and debris, and eventually the death count was established at 450, including 42 schoolchildren.

Anyone who takes water for granted is making a big mistake.

---

came up with a plan to take over management of one of the vital river basin areas that supplies water to a vast amount of the agricultural land. One person compared it to Washington taking over the Mississippi River.

The plan was to buy out farmers from areas with too little water and upgrade irrigation systems that had leaking pipes where water is lost. The prime minister urged citizens to "pray for rain" and told farmers in the Murray-Darling basin area that they would receive no irrigation water without higher inflows into the rivers in the lead-up to winter. The river basin, the size of France and Spain, accounts for 41 percent of Australia's agriculture, 90 percent of the country's irrigated crops, and $22 billion worth of agricultural exports. So dry is the area that surface water is badly needed, but Australia also needs to replenish groundwater systems (aquifers).

While late 2007 and 2008 saw more rainfall, there is no real hope that it will last. Climate change experts indicate that areas such as Southeast Asia may actually become wetter because of global warming, but areas like the interior of Australia will likely become dryer.

Water restrictions are currently in place in many regions and cities of Australia in response to chronic shortages resulting from drought. Depending upon the location, these can include restrictions on watering lawns, using sprinkler systems, washing vehicles, hosing in paved areas, and refilling swimming pools, among others. Increasing population and evidence of drying climates, coupled with corresponding reductions in the supply of drinking water, have led various state governments to consider alternative water sources to supplement existing sources, and to implement "water inspectors" who can issue penalties to those who waste water.

Australia is not alone. In the United States, farmers and government representatives fight over how to keep dwindling water resources like the underground aquifers from disappearing, and in China, Beijing faces a crisis of its own because of terrible pollution affecting the Yellow River. The great river that was once the cradle of Chinese civilization is now terribly polluted and may eventually dry up.

A Water Crisis Catches a Colorado Town by Surprise

The water we take for granted every day when we turn on the tap comes with no guarantees, and this was certainly the case in Alamosa, Colorado, on March 19, 2008, when the Colorado Department of Public Health and Environment issued a bottled-water order for the city and a city/county emergency was declared because of salmonella contamination of the pub-

A Water-Wise Garden

More and more communities are urging residents to plant water-wise gardens, which are low maintenance and low cost and require minimal water. If people select plants suited to their climate, soils, drainage, rainfall, and temperatures, this is an achievable goal. And if homeowners are able to keep the water that naturally falls on gardens from evaporating, supplemental watering would be greatly reduced, which would also decrease water bills. Some areas refer to this as "xeriscaping"—a term coined from "xeros" (Greek for "dry") and "landscaping." Here are some of the methods that help a garden retain water:

- Use mulch
- Increase organic matter
- Collect rainfall
- Use drip irrigation

By using organic mulches, you can increase the amount of organic matter in your soil at the same time. One of the best organic mulches is compost. If you don't have the room to build a compost pile, you can use your grass clippings, leaves, and kitchen waste as mulch. Grass clippings are an excellent source of nitrogen and will last several weeks in your garden. If you can shred and save your leaves from the previous fall, you can incorporate them into your mulch by either mixing them with grass clippings or layering them. Kitchen waste can be put down first and then covered with your mulch. Kitchen waste should only be vegetable matter, as meat waste will attract pests. (When collecting grass clippings to use as mulch, make sure they are herbicide free. The chemicals that are released as the clippings break down can damage your garden plants. If you collect clippings from your neighbors, be sure to ask if they use herbicides. Who knows, by offering to share your harvest in exchange for their clippings, you may convince them to go organic also.)

Collecting rainwater is an easy way to reduce your dependence on your water hose. Divert a downspout into a series of fifty-five-gallon drums to use as needed. Make sure the drums you use did not contain dangerous chemicals and are thoroughly rinsed before you use them. Paint them to match your house or hide them with strategically placed landscaping. If you can locate your rain drums close to your garden, you can run a drip irrigation system from the drums to your garden to deliver the rainwater as needed. A soaker hose made from recycled tires is perfect for this system. You can run sections of regular hose to where you need the water and run soaker hose along the plants.

Using any one of these methods will make an impact on reducing the amount of water your garden needs. Taken together, you can go the entire growing season without having to turn on your water hose in all but the driest years.

lic water supply. Approximately three hundred people were sickened by exposure. Area residents were told to stop using tap water for the time being and only use what was being trucked in for their use.

In the process of trying to root out the problem, the only solution was a flushing of the entire system with massive amounts of chlorine. This, too, presented health risks. With the massive amount of chlorine needed to rid the system of salmonella, health problems could be caused by exposure to the chlorine. Though terrorism and disgruntled employees were ruled out, the cause has remained unidentified. Though the current problem has been resolved, it does show how very vulnerable our water supply actually is.

✪ ✪ ✪

The American Field Service (AFS) was started by World War II ambulance drivers in 1947 with the goal of creating understanding by establishing an environment that could lead to knowledge and familiarity among people from various cultures. The reasoning went that if teens could become friends with people from other backgrounds, this would lead to international cultural understanding.

Perhaps the continuation of this goal is that we learn to respect our neighbors around the world, and that we remember that as technology and easy travel make the world seem smaller, all countries and all citizens need to work together to preserve the geography we've inherited.

Index